"Can spiritual formation take place through disembodied online education? *Ecologies of Faith* answers this question by developing a biblically informed approach to interconnectivity of human relationships that foster spiritual formation. Building on research in the social sciences, the authors illustrate that digital communities can provide an ecology of spiritual connection and growth. This book is a must-read for educators and administrators who are concerned about Christian formation in a digital age."

Mark A. Maddix, dean of the School of Theology and Christian Ministry, Point Loma Nazarene University, San Diego

"When I began teaching seminary in 1996, we offered no online courses or programs; today nearly half of our student course load is online. *Ecologies of Faith in a Digital Age* provides a much-needed theological framework for those of us trying to understand spirituality in our 'brave new *digital* world.' I have known Steve and Mary Lowe for about twenty-five years and have watched their thinking blossom and reach maturity; I look forward to learning much from them about the ecology of spiritual formation!"

R. J. Gore Jr., professor of systematic theology and ministry and dean of Erskine Theological Seminary

"Online and other forms of distance education are growing rapidly in Christian higher education and leaders and teachers in these settings rightly wonder about the opportunities or loss of opportunities for spiritual formation in these formats. Stephen and Mary Lowe provide us with a thoughtful and challenging 'ecological' model of spiritual growth to consider, exploring its presence in Scripture and potential benefit for thinking about and carrying out online teaching in ways that can promote reciprocal spiritual growth as we are mutually connected with Christ. This book is helping me think in new ways about how we can grow together spiritually, even when we are apart, and it has potential benefit for both Christian higher education and mission and local ministry settings. If you are involved in online teaching or ministry, it's a must-read."

Kevin E. Lawson, professor of educational studies, editor of *Christian Education Journal,* Talbot School of Theology, Biola University

"Mary Lowe and Stephen Lowe have long been trailblazers in the work of online theological education. In this volume they bring together their passion for spiritual formation, faithful pedagogy, and online learning. Those who doubt if these things can be brought together will find a well-informed, illuminating, and creative apologetic for faith formation in a digital age. Theological educators and students will be challenged, helped, and encouraged by the work of Mary Lowe and Stephen Lowe. It is a privilege to recommend this fine volume."

David S. Dockery, president of Trinity Evangelical Divinity School

"Theological education is so much bigger than what happens in the classroom. In our increasingly interconnected world we're just now beginning to realize how much that truth impacts every aspect of teaching and learning. Nowhere is that more important than online. This rich resource helps distance educators think about their work less as information download and more as sowers of seed, cultivating and nurturing both soil and plant in a holistic ecosystem of transformative learning."

Mindi Thompson, associate professor and director of distance education at Abilene Christian University's Graduate School of Theology, president of the ACCESS Association of Christian Distance Educators

"This is a thorough and insightful depiction of the potential for online learning to promote significant transformation in the lives of your learners. If your institution is looking into the future of spiritual formation in digital worlds, this book will be instrumental in guiding you."

Lorne Oke, executive director at the Center for Learning and Innovation, Indiana Wesleyan University

"In this ubiquitously connected and networked age, Steve Lowe and Mary Lowe have offered a thought provoking, ecological perspective of spiritual formation and how we grow as Christians. Their ecology-of-faith model can provide keen insight to providers of Christian, online education and may contribute to meaningful conversations regarding how the mission of Christian education aligns with online delivery. I believe this is a must-read!"

Kaye Shelton, associate professor of educational leadership at Lamar University

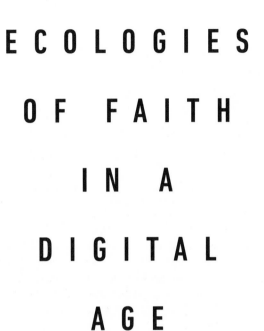

ECOLOGIES

OF FAITH

IN A

DIGITAL

AGE

SPIRITUAL GROWTH
THROUGH
ONLINE EDUCATION

STEPHEN D. LOWE AND MARY E. LOWE

IVP Academic

An imprint of InterVarsity Press
Downers Grove, Illinois

InterVarsity Press
P.O. Box 1400, Downers Grove, IL 60515-1426
ivpress.com
email@ivpress.com

InterVarsity Press® is the book-publishing division of InterVarsity Christian Fellowship/USA®, a movement of students and faculty active on campus at hundreds of universities, colleges, and schools of nursing in the United States of America, and a member movement of the International Fellowship of Evangelical Students. For information about local and regional activities, visit intervarsity.org.

All Scripture quotations, unless otherwise indicated, are taken from the New American Standard Bible®, copyright 1960, 1962, 1963, 1968, 1971, 1972, 1973, 1975, 1977, 1995 by The Lockman Foundation. Used by permission.

Figure 2: John Muir in pear orchard at home, Martinez, California. Photo by George R. King (ca. 1912). John Muir Papers, Holt-Atherton Special Collections, University of the Pacific Library. © 1984 Muir-Hanna Trust. Used by permission.

Cover design: Tim Green / Faceout Studio
Interior design: Daniel van Loon
Images: hand illustration: © saemilee / Digital Vision Vectors / Getty Images
        circuit board: © watchara / Shutterstock.com
        vine: © Lawrence Del Mundo / Stocksy.com

ISBN 978-0-8308-5205-5 (print)
ISBN 978-0-8308-8743-9 (digital)

Printed in the United States of America ♾

InterVarsity Press is committed to ecological stewardship and to the conservation of natural resources in all our operations. This book was printed using sustainably sourced paper.

Library of Congress Cataloging-in-Publication Data
Names: Lowe, Stephen D., 1949- author.
Title: Ecologies of faith in a digital age : spiritual growth through online
    education / Stephen D. Lowe and Mary E. Lowe.
Description: Downers Grove : InterVarsity Press, 2018.
Identifiers: LCCN 2018012236 (print) | LCCN 2018015490 (ebook) | ISBN
    9780830887439 (eBook) | ISBN 9780830852055 (pbk. : alk. paper)
Subjects: LCSH: Spiritual formation. | Christian education. |
    Education--Effect of technological innovations on. | Internet in education.
Classification: LCC BV4511 (ebook) | LCC BV4511 .L69 2018 (print) | DDC
    230.078/5--dc23
LC record available at https://lccn.loc.gov/2018012236

P 24 23 22 21 20 19 18 17 16 15 14 13 12 11 10 9 8 7 6 5 4 3 2 1

Y 38 37 36 35 34 33 32 31 30 29 28 27 26 25 24 23 22 21 20 19 18

To all of our colleagues

in online Christian education

around the world.

CONTENTS

Acknowledgments   ix

Introduction   1

PART 1: A BIBLICAL THEOLOGY OF ECOLOGY

1 God's Garden Ecology   11

2 The Ecological Motif in Scripture   24

3 The Ecological Power of Parables   38

4 The Ecology of the Body of Christ   50

PART 2: SPIRITUAL FORMATION THROUGH DIGITAL ECOLOGIES

5 Growing Together Online   67

6 Designing Learning Ecologies   87

7 Digital Ecologies Across the Digital Landscape   104

8 Social Networks and the Power of Reciprocal Influence   120

PART 3: ECOLOGICAL CONNECTIONS TO CHRIST AND COMMUNITY

9 Ecological Connection to Christ   137

10 Ecological Connections to Christians   150

11 Ecological Interactions with Other Christians   171

12 Ecological Sanctification   188

13 Thinking Ecologically About Spiritual Growth   207

Conclusion   227

Bibliography   229

Author Index   247

Subject Index   251

Scripture Index   257

ACKNOWLEDGMENTS

This book has been fermenting at different stages over a number of years. The original idea for the concept of an ecology of spiritual formation began with a course ("The Moral Aspects of Spiritual Development") Steve took at Princeton Theological Seminary with his doctoral dissertation chair at Michigan State University, the late Dr. Ted Ward. In that course, Ted encouraged the students to consider spiritual formation from an ecological perspective and recommended the work of Urie Bronfenbrenner as a way of thinking about this novel approach to conceptualizing how we grow as Christians.

Around 2007, Mary completed her doctoral studies with a dissertation that focused on assessing spiritual formation in theological distance education. This work provided the research basis for a grant we received from the Wabash Center for Teaching and Learning in Theology and Religion that enabled us to cohost a three-year National Consultation on Spiritual Formation in Theological Distance Education. We continue to be grateful to the participants in that group of pioneering online educators who helped refine and improve our work. It was during this consultation that we introduced the ecologies of faith concept that we had been using in a course on spiritual formation at Erskine Theological Seminary. That consultation launched a series of peer-reviewed articles addressing different aspects of our model that garnered interest from online practitioners around the world.

The initial step on the journey to publication came through a conversation with Joanne Jung at Biola University, who then put us in contact with David McNutt, associate editor at IVP Academic. David has provided us with careful, thoughtful, and encouraging direction throughout the process, and for this we are deeply grateful.

We would be remiss if we did not acknowledge the encouragement and friendship of our colleagues at Erskine Theological Seminary and our Outreach Sunday school class, both of whom listened patiently and attentively as we explained our concept of spiritual formation and tried as best we could to answer their thoughtful and probing questions. We also wish to express deep appreciation to former students at Erskine Theological Seminary and current students in the Rawlings School of Divinity at Liberty University, who actively participated in our courses where we taught many of the concepts set forth in the following chapters.

We also wish to thank Kathrin Herr for her careful editorial eye in the early stages of manuscript development and in her work on our bibliography and index.

We are both products of parents who served the Lord faithfully in different spheres of the kingdom. Steve's parents, Rev. Alfred and Helen Lowe, served in pastoral ministry for almost fifty years before the Lord called them home. Mary's parents, Don and Evelyn Adams, served as church-planting missionaries in Haiti and Suriname for over forty years. We are grateful for these godly parents who instilled in us a love for God's word and God's people. We are also thankful for our children, grandchildren, and extended family members who always asked, "When is your book going to be done?" Well, here it is.

INTRODUCTION

For even though I am absent in body, nevertheless I am
with you in spirit, rejoicing to see your good discipline
and the stability of your faith in Christ.

COLOSSIANS 2:5

The experience of teaching online courses and guiding doctoral research online evolved over the years for us. We did not set out to teach online courses or write about issues related to online Christian education. Expectations for our professional academic careers involved a beautiful campus with ivy-covered walls and a close-knit academic community of colleagues and students. While we certainly have had that, we have also had an unexpected blessing given to us through our participation in online Christian education. We believed, maybe like you, that online learning was inferior to traditional education on a physical campus and in a typical classroom setting. This was especially true in our early days of distance education using VHS videotapes, audio cassettes, postal delivery of course materials, and mediated communication through keyboarding in real time. We can still remember our first ventures in synchronous live "webinars" with students from all across the United States who met in real time for about an hour of discussion and interaction. The poor students who could not type fast had a hard time keeping up with the pace of the discussion and exchange of messages through CompuServe.

With the advances in learning technologies, many of the old distance barriers and the clunky means of communication have vanished. Just recently, Steve conducted a research methods class for doctoral students in

South Korea while seated in a video classroom in Virginia. The quality of the images and the ease of communication between professor and students in real time was startling. Students could see his PowerPoint slides as if they were in the room with him, and he could see the expressions on their faces and what they were eating for snacks during the breaks. A host of other technological innovations that many who teach online employ on a regular basis makes using such technology seamless and unobtrusive.

Two of the major stumbling blocks for many who resisted teaching online were community formation and spiritual growth. We understood the concerns because we had to face these issues ourselves as we migrated over the years to online teaching. While each of us had our educational, sociological, and theological reasons why both are problematic or potentially possible in online settings, the feedback and unsolicited comments from our students have sealed our conclusions about both community formation and spiritual growth in online courses and degree programs.

One student, who was taking an online course for the very first time, wrote in his online journal, "[As a result of taking this online class,] I can report that I have grown not only academically but in every area of my life. That is to say, I have grown holistically as a person and as a child of God. My personal goals have changed and evolved throughout the journey of my studies." A doctoral student in an online degree program wrote, "Not only am I allowed to pursue my chosen degree, I am also able to connect to a network of people whom I would have never met, much less interacted with and formed a community." Another online student provided this unsolicited comment, "I have experienced the most accelerated spiritual growth while completing my master of divinity online and it was because of the collaborative learning process and sense of community. I was able to learn deep, theological truths and how they apply to my life through interacting with classmates, professors, and the course material. In turn, my social network (friends, family, church, neighbors, etc.) benefitted, as I would share what I was learning with them through discussions, Sunday school, and Facebook." Space does not permit including other written feedback we have received over the years from both students and faculty. Those who have made the decision to participate in online Christian education have discovered that many of the prospective drawbacks and concerns vanished as a result of their experience, and they became enthusiastic supporters of this form of educational delivery. Many have been pleasantly surprised at the rich and thick nature of the interactions and exchanges that permit a more thoughtful and nuanced consideration of an issue or topic. While there is certainly a great deal of room for improvement and quality enhancement of the online experience, those who have experienced it firsthand feel part of an authentic, Christian community and sense a positive growth in their walk of faith as a result of their virtual encounter with other Christians. These experiences in online Christian education happened because we have entered a new era of unprecedented technological innovation that makes possible the creation of spiritual connections between believers—connections that have never existed before in human history.

AN INTERCONNECTED AGE

Alvin Toffler (*The Third Wave*, 1984) gave us his take on the different eras of human history in developed countries. He identified three waves that divide human history into four ages: the nomadic age, the agricultural age, the industrial age, and finally, the information age. While Toffler has never proposed a fourth wave or fifth age, we argue that we are now on the cusp of the interconnected age. This is an emerging age that is transitioning from the information age and providing us with multiple ways to interconnect with people, places, ideas, religions, institutions, organizations, and affinity groups. The physical sciences help us see an interconnected universe "where every atom is connected to every other atom."[1] Another way of describing the interconnected age is to see it as a "networked world" or "the age of the network."[2] The interconnected age would not be possible without the computer, the internet, and a host of digital devices at our disposal.

Heidi Campbell and Stephen Garner argue that the church needs to ponder how it will negotiate faith given the interconnected age in which it now exists.[3] The network society in which the church currently lives, they

---

[1]Brian Cox and Jeff Forshaw, *The Quantum Universe* (Boston: Da Capo Press, 2011), 140.
[2]Lee Rainie and Barry Wellman, *Networked: The New Social Operating System* (Cambridge, MA: MIT Press, 2014), 255; Jessica Lipnack and Jeffrey Stamps, *The Age of the Network: Organizing Principles for the 21st Century* (Essex Junction, VT: Oliver Wright, 1994).
[3]Heidi Campbell and Stephen Garner, *Networked Theology: Negotiating Faith in a Digital Culture* (Grand Rapids: Baker Academic, 2016).

argue, requires the church to ask about the effects on our theologizing and spiritual lives. What is the relationship between new media and theology? We argue, similarly, that the church and its institutions and organizations need to consider the impact of the interconnected age or network society on our concepts, paradigms, and practices of spiritual formation.

We have seen evidence in the church, classroom, and community that traditional approaches to spiritual formation are inadequate because they are too individualistic, private, and disconnected. Most of us no longer live in monasteries cloistered from the real world. All of us live, work, and relate to each other in the real world and this world is digital, networked, and highly interconnected. Given the twenty-first century reality of digital relationships and interactions, we need a model of spiritual formation that provides a better fit for the way we live in the interconnected age.

AN ECOLOGICAL MODEL OF SPIRITUAL FORMATION

Since we live in a highly interconnected world where first-year students know about their roommates before they ever arrive on campus and where we may interact with people overseas as much as we do our neighbors next door, we propose a model of spiritual formation that recognizes and utilizes these interconnective and interactive realities. In the natural world nothing grows alone, isolated and disconnected from its ecological habitat. Instead, everything grows ecologically through connections to and interactions with other living and nonliving things producing mutual growth and fecundity. We will argue that there is an ecological motif running through the Bible that uses the ecological growth observed in nature to illustrate spiritual growth. Jesus taught his parables upon the assumption that what is valid in one sphere (the natural) is valid also in the other (the spiritual).

An ecological perspective on how we grow as Christians enlarges our orientation to spiritual formation and engenders relationships and connections beyond our traditional privatized perceptions of how Christians grow. An ecosystem perspective reminds us, indeed requires us, to think of growth both individually and corporately. Defined natural ecologies like forests, gardens, vineyards, and fields, as well as defined social ecologies like families, workplace, and educational institutions, inform us about how spiritual ecologies like the church, Christian homes, Christian organizations, and other faith communities function to promote spiritual growth. In God's ecology, individual things and people do not grow alone. They grow when they connect to and interact with the ultimate Source of Life and other growing people.

A ROAD MAP THROUGH ECOLOGIES OF FAITH

We think ecological images and metaphors not only appear in Paul's description of the body of Christ but also permeate the entire landscape of God's revelation in Scripture. In chapter one we see that God created natural and human ecologies beginning with the perfect ecology of the Garden of Eden. Although this was a perfect place for human and biological growth, the introduction of human sin marred God's perfect ecology and introduced disconnection at multiple levels of existence. These multiple disconnections require reconciliation that can only come through the reconciling work of Christ on the cross.

Chapter two introduces the concept of an ecological motif that runs from Genesis to Revelation. After identifying the chief characteristics of a biblical motif, we provide examples from both the Old and New Testaments of an ecological motif in which the writers of Scripture use examples or illustrations from natural ecology to illustrate spiritual patterns of growth. We show how the Bible references trees, flowers, seeds, gardens, vineyards, forests, olive groves, and fields to identify different features that suggest growth and expansion.

In chapter three, we examine the ecological power of the nature parables that Jesus taught. We examine the work of several parable scholars who provide evidence for another feature of the ecological motif in Scripture. In particular, we observe that in the parables Jesus taught, there is a direct correspondence between how things grow in nature and how they grow in the kingdom of God. We examine selected parables that utilize observations from natural growth to illustrate the processes of spiritual growth. We then draw some applications from this biblical evidence for understanding Christian spiritual formation.

The focus of chapter four is a study of the ecological overtones of the human body metaphor Paul applies to the body of Christ. In particular, we highlight the work of Howard Snyder, who most consistently argues for thinking about the church through an ecological lens. We draw parallels between the ecology of the human body and the ecology of the body of Christ, showing how growth occurs through a series of spiritual connections and interactions.

Beginning in chapter five, we examine the online learning ecology and show how our model of spiritual formation fits this newly emerging form of education. Online or any form of distributed education beyond the campus classroom requires a model of learning that recognizes and appreciates the distinctive nature of the online environment. We offer evidence that online Christian communities of learning are just as valid and authentic as on-campus communities and can produce the same spiritual outcomes. In chapter six, we consider the design and structure of online learning ecologies. We are concerned to show how social connections and learning interactions form an online learning ecology that creates a learning web of interconnected participants (students and faculty), concepts (forming a logical structure of course content), course materials, learning tasks, and educational resources. Chapter seven provides an opportunity for us to investigate the digital ecologies that allow us to interconnect with each other using a variety of mediated technologies. Critics have charged that online experiences are disembodied and thus incapable of forming community and facilitating spiritual growth. We argue from the example of the apostle Paul that any form of mediated communication (letters, computers, digital devices) may convey personal presence and promote spiritual growth. In chapter eight, we direct our attention to the power of social networks and what they can teach us about how Christians grow ecologically through social and spiritual connections and interactions. We highlight social network research that demonstrates the power we have to influence mutually those with whom we have some social bond. We draw parallels between the social connections and mutual social interactions we have in social networks to the spiritual connections and mutual spiritual interactions we have in the spiritual ecology of the church.

With chapter nine, we explore an exegetical analysis of Pauline terminology that seems to fit the ecological paradigm. In chapter nine, we examine Paul's use of what we identify as "vertical *syn*-compounds" wherein he establishes our spiritual connection to the person and work of Christ. The need for establishing a connection to Christ finds it origin in the chapter on the Garden ecology and the introduction of human sin that creates multiple forms of disconnection in the universe. Through the reconciling work of Christ on the cross, we enjoy a spiritual connection to Christ that makes possible our fellowship with God. Chapter ten examines the same *syn*-compound terms but now used in reference to connected members of the body of Christ. This is what we term "horizontal *syn*-compounds" because Paul uses them to stress how connected we are to one another as members of Christ's body. In chapter eleven, we build upon the connectional language with a study of Paul's use of the *allēlōn* ("one another") reciprocal imperative. In this chapter, we study how Paul applies this language to the interactions, relationships, exchanges, and various transactions that occur between and among believers within the church. Just like Urie Bronfenbrenner demonstrated through his ecology of human development model that mutual social interactions lead to mutual development in human growth, so too Paul's use of this language demonstrates that mutual spiritual interactions lead to mutual development in spiritual growth.

Chapter twelve sets out an ecological take on sanctification and holiness that complements traditional instruction on these topics. From his public ministry, we argue that Jesus understood holiness and sanctification to be contagious qualities that spread from one person to another. We argue from Rodney Stark's sociological analysis that the contagion of the gospel spread through Christian social networks and made possible the gradual growth of the Christian faith within the Roman Empire. We also show from Ephesians 4 that individual holiness spreads from one member of the body of Christ to another, thereby making it possible for the body to build itself up spiritually.

We conclude the book in chapter thirteen with an exhaustive item-by-item summary of the salient features of our ecologies of faith model. In this way, we recapitulate the major themes and topics in a way that cements comprehension and understanding. Our purpose in all of this is to set out a biblically grounded model of spiritual formation that integrates insights from sociology, ecology, developmental psychology, education, and the emerging field of social network theory and apply it to online Christian education. Although our field of vision is limited to that particular area of ministry praxis, we recognize that the model may have wider application to other sectors of Christian ministry. It is our hope and prayer that those who serve in other areas of ministry will creatively apply our proposal and find it beneficial and effective.

A PERSONAL INVITATION

We are both administrators and teaching faculty on a large university campus teaching courses in traditional classrooms. We value this aspect of our academic and professional career and realize that the embodied experience of teaching and learning has a rightful place in any ecology of spiritual formation. However, the question about where spiritual growth occurs needs a more holistic and ecologically informed perspective if we are to capture the essence of biblical teaching and apply it to various forms of Christian community whether digital, virtual, or physical. The ecology-of-faith model we propose in the following chapters offers a promising start. It blends a variety of disciplinary elements into a comprehensive and holistic framework that encourages us to think outside the box of traditional formulations that may no longer serve Christian ministry well in the interconnected age. We invite you to join us on this journey of mutual discovery. We hope that you are encouraged and challenged to grow—even though we may not be present in body with you.

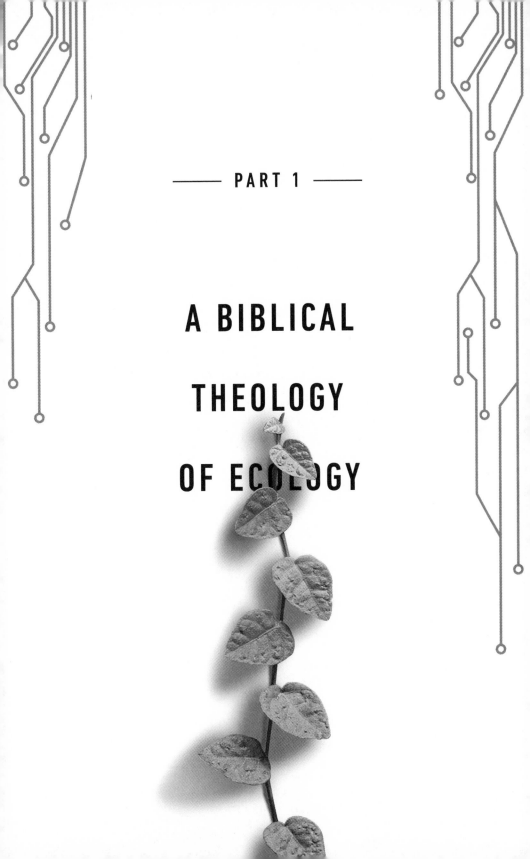

—— PART 1 ——

A BIBLICAL

THEOLOGY

OF ECOLOGY

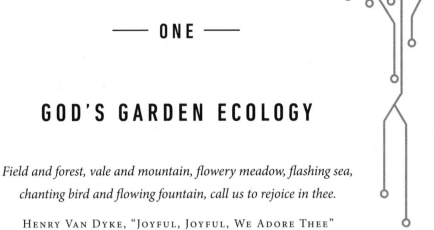

── ONE ──

GOD'S GARDEN ECOLOGY

Field and forest, vale and mountain, flowery meadow, flashing sea,
chanting bird and flowing fountain, call us to rejoice in thee.

HENRY VAN DYKE, "JOYFUL, JOYFUL, WE ADORE THEE"

One of the most stunning synagogue mosaics unearthed outside of Israel is the one on the floor of the synagogue in the ancient Cilician city of Mopsuestia in modern-day Turkey. The original builders paved the entire synagogue in mosaics depicting Noah's ark surrounded by a variety of animals and the story of Samson. These mosaics and others like them illustrate the mastery required to take separate tiles and form them into a coherent scene that looks as if they were part of an original whole. In this synagogue, the original artisans were able to synthesize two different biblical accounts into one in a seamless unity that is breathtaking to behold.[1]

When we study God's creative masterpiece, we discover that we live in a physically interconnected universe. Biological, social, and physical scientists all agree that we live in a universe more akin to a spider's web or power grid than to telegraph or telephone lines strung between poles. Ervin Laszlo summarizes the consensus view: "Most scientists are ready to admit that, in principle, 'everything is connected with everything else.'"[2] The interconnected reality of the universe is supported by quantum physicists when they assert that "every atom in the Universe is connected to every other atom."[3]

---

[1]Lee I. Levine, *Ancient Synagogues Revealed* (Jerusalem: The Israel Exploration Society, 1981), 187.
[2]Ervin Laszlo, *The Interconnected Universe: The Conceptual Foundations of Transdisciplinary Unified Theory* (Singapore: World Scientific Publishing, 1995), 9.
[3]Brian Cox and Jeff Forshaw, *The Quantum Universe* (Boston: Da Capo Press, 2011), 140.

We now understand that nothing in the universe happens in isolation. We live in a highly interconnected, spider-like web of interlocking elements at every level of human existence (biological, economic, electronic, digital, political, molecular, sociological, physical, and spiritual). All this evidence tells us that God created a world where things grow, develop, and are sustained through ecological connections and interactions. Our task as God's servants is to recognize this created reality and live our lives, grow spiritually, and teach and preach in harmony with it. In this chapter, we will consider God's creative masterpiece from an ecological orientation. We will first zoom out upon God's cosmic ecology and then zoom in on the Garden ecology depicted in the opening chapters of Genesis.

COSMIC ECOLOGY

As we embark on our consideration of an ecological model, we need to start by pulling the camera back and widening our view. Edward Wimberley provides such a perspective when he writes about "cosmic ecology."[4] Rather than thinking only about what he calls "environmental ecology," which is centered on the earth, he situates earth's ecology within the larger frame of the ecosystem of the cosmos: "While suspended in the vacuum of the universe, the planet Earth remains inextricably embedded within a cosmic ecosystem that sustains it and largely dictates the conditions under which life on the planet will exist."[5] He cites specific examples of this cosmic connection by reference to "cosmic rays produced by distant exploding stars [that] interact with airborne particles in the lower atmosphere to create heavier cloud formations, producing global shading, rain, and cooler temperatures."[6]

He argues that our current notions of environmentalism and environmental ecology are too restrictive because they fail to appreciate what he calls "our cosmic 'household,'" a phrase that originated from the Greek word *oikos*.[7] In his view, the known universe functions "as a set of interrelated ecosystems that are of great value and importance."[8] He proposes

---

[4]Edward T. Wimberley, *Nested Ecology: The Place of Humans in the Ecological Hierarchy* (Baltimore: Johns Hopkins University Press, 2009), 4.
[5]Ibid., 67.
[6]Ibid.
[7]Ibid., 77.
[8]Ibid., 73.

thinking of the existing universe as a "nested ecology" consisting of "an interlocking set of systems that begin at the level of the self and progressively extend to encompass families, groups, communities, ecosystems, the biosphere, and beyond into the unfathomable reaches of the cosmos."[9] According to Wimberley, human beings nest within existing natural ecologies but "are also embedded within an intricate network of personal and social ecologies that are predicated upon the existence of a vast and complex web of natural ecologies."[10] Our entire existence as humans consists of being intricately interconnected to and interacting with all of the various overlapping and interdependent ecological entities, some of which we will delineate in passing before we embark upon a consideration of the ecology of the Garden of Eden.

EARTH ECOLOGY

A natural ecosystem—one that we find in God's creation—is a specific environment in which living things grow. Biotic (living) and abiotic (nonliving) components work in a mutual manner to facilitate growth and capacity to reproduce. Without ecological connections and interactions, there is no growth and reproduction. To grow, living things need ecologies and specific habitats within those ecologies.

The English words *ecology* and *ecosystem* derive from the Greek word *oikos* and other words in its semantic domain.[11] There are approximately three hundred occurrences of *oikos/oikia* and the larger semantic field in the New Testament. John H. Elliott writes, "Altogether this constellation of terms forms one of the largest semantic fields in the entire NT."[12] The Greek Old Testament (Septuagint) translates *oikos* from the Hebrew word for

---

[9]Ibid., 4.

[10]Ibid., 3.

[11]Johannes P. Louw and Eugene A. Nida, *Greek-English Lexicon of the New Testament Based on Semantic Domains* (New York: United Bible Societies, 1989) lists *oikeios* (relative), *oiketeia* (household servants), *oiketēs* (house servant), *oikeō* (dwell), *oikēma* (quarters), *oikētērion* (dwelling place), *oikia* (house, family), *oikiakos* (relative), *oikodespoteō* (manage a household), *oikodespotēs* (manager of household), *oikodomeō* (build a house), *oikodomē* (building), *oikodomos* (builder), *oikonomeō* (manage a household), *oikonomía* (household management), *oikonomos* (administrator of household), *oikoumenē* (habitation of the earth), *oikourgos* (homemaker).

[12]John H. Elliott, "Household/Family in the Gospel of Mark as a Core Symbol of Community," in *Fabrics of Discourse: Essays in Honor of Vernon K. Robbins*, ed. David B. Gowler, L. Gregory Bloomquist, and Duane F. Watson (Harrisburg, PA: Trinity Press International, 2003), 41.

family or household (*bayit*). German zoologist Ernst Haeckel first coined the term *ecology* in 1866 and defined it as "the science of relations between the organism and the surrounding outer world."[13] Alfred George Tansley first introduced the term *ecosystem* to scientific literature in his 1935 article "The Use and Abuse of Vegetational Concepts and Terms."[14] The word *oikos* originally referred to any habitation or space where people, animals, and all living things resided. J. Donald Hughes writes that *oikos* was the word literally used to refer to "the whole inhabited earth . . . the home of all mankind."[15] Collomb reminds us, "In ecology nature is indeed seen as *oikos* . . . a biosphere made up of interrelated parts."[16] These descriptions of ecology as *oikos* honor Haeckel's reference to ecology as the "household of nature."[17] In addition, *oikos* referred to a house, temple, household, apartment, or various humans who inhabited such structures.

Various and numerous definitions are extant in the ecological literature but no one recognizes any of them as definitive.[18] The simplest definition of ecology or ecosystem we have states that it is the study of the "interdependence of all living things, plant and animal, and their environment which makes life possible."[19] Although the scientific community uses ecology and ecosystem distinctively and analytically, we see ecology and ecosystem for our purposes as synonyms and will often use them interchangeably.

Although the study of ecology existed primarily in the fields of biology and zoology and originally focused on the relationship between organisms and their environment, eventually the concept broadened to other natural and social sciences as a way to explain the complex relationships between

---

[13]Ernst Haeckel originally coined the term in *Generelle Morphologie der Organismen* (1866), but it appeared later in W. C. Allee, *Principles of Animal Ecology* (Philadelphia: W. B. Saunders, 1949); Fritjof Capra, *The Web of Life: A New Scientific Understanding of Living Systems* (New York: Anchor, 1996), 33.

[14]Alfred George Tansley, "The Use and Abuse of Vegetational Concepts and Terms," *Ecology* 16, no. 3 (1935): 284-307.

[15]J. Donald Hughes, *Ecology in Ancient Civilizations* (Santa Fe: University of New Mexico Press, 1975), 3.

[16]Jean-Daniel Collomb, "Christianity to Ecology: John Muir's Walk Through America," *Transtext(e)s Transcultures* 4 (2008): 108, http://transtexts.revues.org/254, doi:10.4000/transtexts.254.

[17]K. Friederichs, "A Definition of Ecology and Some Thoughts About Basic Concepts," *Ecology* 39, no. 1 (1958): 154.

[18]Robert Leo Smith, *Ecology and Biology* (Menlo Park, IL: Addison Wesley Longman, 1996).

[19]Edward J. Ryan, "Ecology: A Study of Interdependence," *The Kansas School Naturalist* 11, no. 1 (1964): 3.

interconnected entities, such as the fields of family, congregational, social, human, and economic ecologies. We will see in subsequent chapters how this same conceptual perspective applies to our understanding of the church, the kingdom, and Christian spiritual growth.

SOCIAL ECOLOGY

Ecology does not only refer to various biomes of the created world. All human beings exist within a defined group of people with whom we regularly connect and interact with in some fashion. As Peter Berger states, "Man is biologically predestined to construct and to inhabit a world with others."[20] We would prefer to say that humans are *divinely* predestined to construct and to inhabit a world with others. God built this social dimension into our creational DNA—made clear in the book of Genesis: "Then the LORD God said, 'It is not good for the man to be alone; I will make him a helper suitable for him'" (Gen 2:18). The image of God has something to do not only with our relationship to God but with our relationship to others as well. The two key words in social ecology are the same key words that surface in any ecology: *connections* and *interactions*. Connections form some kind of a social bond, usually embodied in a social organization (family, business, church, organization, institution). By comparison, interactions refer to what transpires between the connections or the social capital that emerges from the interactions.

The emerging field of social network analysis sharpens our understanding of social ecologies. One of the main features of our social ecologies is the concept of "social contagion." Social contagion has to do with our ability to influence others and the ways in which others influence us. Nicholas A. Christakis and James H. Fowler's research on social networks shows six degrees of separation that explain how connected we are to one another, while three degrees of separation explain how contagious we are to one another.[21] The dwarfs in *Snow White and the Seven Dwarfs* conveniently illustrate six degrees of separation. Between each of the seven dwarfs is a degree of social separation

---

[20]Peter L. Berger, *The Sacred Canopy: Elements of a Sociological Theory of Religion* (New York: Anchor, 1967), 183.

[21]Nicholas A. Christakis and James H. Fowler, *Connected: The Surprising Power of Our Social Networks and How They Shape Our Lives* (New York: Little, Brown, 2009).

totaling six degrees of separation. Three degrees of separation would be the friend of a friend of a friend. Christakis and Fowler note this phenomenon operating in social ecologies: "We need to understand how interconnections and interactions between people give rise to wholly new aspects of human experience that are not present in the individual themselves."[22] The social connections we have with other people and the interactions we create to maintain those connections make us socially contagious—for better or worse.

The spiritual ecology of the body of Christ partakes of the same connections and interactions necessary for spiritual growth and influence (spiritual contagion). In subsequent chapters, we will study Paul's use of *syn*-compounds ("together with") that emphasize our spiritual connection to Christ and to other Christians. In addition, we will examine Paul's use of *allēlōn* ("one another"), reciprocal imperatives that stress our spiritual interactions with other Christians to whom we mutually connect through Christ.

HUMAN ECOLOGY

Human growth takes place within human ecologies that represent social connections to and interactions with other developing humans—connections and interactions that lead to mutual human growth. The pioneering and influential work of Urie Bronfenbrenner in *The Ecology of Human Development* guides our study of human ecology. He argues that it is not enough to postulate the existence of innate patterns of development that unfold in a mechanistic manner. He insisted that human growth occurs as the result of bonded interconnections and reciprocal interactions between and among developing persons. He described bonded interconnections as "the relations of the various parties with each other as members of a group engaged in common, complementary, or relatively independent undertakings."[23] He went on to propose that these social bonds create the "phenomenon of interaction [as] fundamental to an understanding of how human beings develop."[24] He used the term *reciprocity* to describe a "phenomenon of interaction" that involved what he described as "concomitant mutual feedback" that generated

---

[22]Ibid., 32.
[23]Urie Bronfenbrenner, *The Ecology of Human Development: Experiments by Nature and Design* (Cambridge, MA: Harvard University Press, 1979), 25.
[24]Ibid., 146.

"progressively more complex patterns of interaction."[25] The result of these reciprocal engagements "produces its most powerful developmental effects" on the mutual growth of all interacting persons.[26]

Bronfenbrenner further proposed that reciprocal interaction between developing persons in human ecologies produces reciprocal development among all active participants. The social connection and interaction between persons serve as "a vehicle with a momentum of its own that stimulates and sustains developmental processes . . . as long as they remain interconnected . . . in a bond."[27] Thus, what Bronfenbrenner observed as occurring in human social ecologies we find consistent with natural and social ecologies.

PERSONAL ECOLOGY

What we observe as characteristic functions of natural, social, and human ecologies we find present in the individual as well. In the foreword to *Nurture That Is Christian*, the late Ted Ward offered what he called "a theory of spiritual ecology," which borrowed the concept of ecology from the study of biology "that refers to the interdependence of each component in the creation with respect to each other component."[28] He used this to guide his concept of how each human being "is internally related," where "every part is joined together in such a way as to affect each other for mutual better or worse."[29] He argued that the study of human development offered a tantalizing way to organize the various aspects of human being into a personal ecology represented—but not exhausted—by six different developmental dimensions: physical, intellectual, emotional, social, moral, and spiritual.

Taking our cue from Ward, we have conceptualized this personal ecology using the human hand to illustrate the "interdependence of each component . . . with respect to each other component" as seen in figure 1.

In the same way that any defined ecology functions through connections and interactions to produce mutual growth, so does the personal ecosystem of the human being. Just as Bronfenbrenner identified the dynamic elements

---

[25]Ibid., 57.
[26]Ibid.
[27]Ibid., 66.
[28]James C. Wilhoit and John M. Dettoni, eds., *Nurture That Is Christian: Developmental Perspectives on Christian Education* (Wheaton, IL: BridgePoint, 1995), 14.
[29]Ibid.

of our human ecologies that interact with one another to produce reciprocal development, so do the interactive dimensions of each human seek to produce whole-person development across all dimensions. Neglect of any one or more of these dimensions, in one way or another, diminishes the ability of the human to reach full maturity and developmental equilibrium. In the same way, Christians who are to "grow up in all *aspects* into Him who

is the head, *even* Christ" must reflect growth in all of these aspects or dimensions of our created personhood if we are to achieve "the measure of the stature which belongs to the fullness of Christ" (Eph 4:15, 13, emphasis added).

GARDEN ECOLOGY

As we construct our view of spiritual formation through the lens of ecology, we must begin by considering the original and perfect ecosystem God created in the Garden

Figure 1. Six developmental dimensions

of Eden. Although we may identify multiple ecologies existing simultaneously in the cosmic ecosystem, the first defined ecology was the work of God on behalf of his animal and human creatures in the Garden of Eden. God's original humans nested perfectly within what Theodore Hiebert called "the ecology of the garden."[30] The garden ecology was a pristine, divinely created environment in which, to quote Paul, "God causes all things to work together for good" (Rom 8:28). The symmetry and symphony of creation functioned perfectly and blessed Adam and Eve, as well as the animals and vegetation. Hiebert demonstrates that "the ecology of this garden is distinct from the Mediterranean highlands" and must be viewed as a unique ecological environment.[31]

---

[30]Theodore Hiebert, *The Yahwist's Landscape: Nature and Religion in Early Israel* (New York: Oxford University Press, 1996), 52.
[31]Ibid.

The oasis ecology of the Garden of Eden is an extended agricultural image of God's original interaction with humanity. God the original farmer "planted a garden" and placed the human in that pristine ecology (Gen 2:8). Once God planted the garden ecology, it became the task of the humans to maintain what the Lord had created—"to cultivate it and keep it" (Gen 2:15). All of the named components of the garden functioned interactively to produce the perfect ecology for humans to enjoy and flourish as God intended.

In Genesis we see the garden described as a "luxuriantly fertile place."[32] For example, Genesis 13:10 says, "All the valley of the Jordan . . . was well watered everywhere . . . like the garden of the LORD, like the land of Egypt as you go to Zoar." The floodwater irrigation of the Nile River in Egypt is a base of comparison with the garden of the Lord. One even sees this rich ecology reflected in the name Eden, which means "lush fecundity."[33] The Genesis 2 account lends support to the Genesis 13 description with the reference to the four rivers that flowed from it to water the whole earth (Gen 2:10-14). The Garden of Eden was the perfect oasis ecology for the humans God created in his image and likeness. This is supported by the Hebrew term "garden" (*gan*), which is used in biblical Hebrew for agricultural practices requiring irrigation. Since the garden ecology existed prior to rainfall (as we read in Gen 2:5), it was watered by irrigation (*šqh*) that found its source in a spring (*'ēd*) that emerged from underground (Gen 2:6). Such a defined garden ecology, with its constant irrigation, temperate climate, and biodiversity, reflected a pristine environment captured in the Semitic root for Eden, *'ēden*, which means "fertility."[34]

God the original farmer "planted a garden" and delegated his farming tasks: "Then the LORD God took the man and put him in the Garden of Eden to cultivate [*'bd*] it and keep it [*šmr*]" (Gen 2:8, 15). Yet regardless of the human activity, the narrative makes it clear that God is the ultimate source of any growth (Gen 2:9). Growth produced through the combined efforts of the human and God takes place within a defined ecosystem of

---

[32]Richard Bauckham, *The Bible and Ecology: Rediscovering the Community of Creation* (Waco, TX: Baylor University Press, 2010), 108.

[33]Bruce K. Waltke, *Genesis: A Commentary* (Grand Rapids: Zondervan, 2001), 85.

[34]Hiebert, *Yahwist's Landscape*, 55.

complementary reciprocity—of giving and receiving nutrients and nourishment. All of the named components of the garden and those not named functioned interactively to produce the perfect ecology for humans to enjoy.

THE GARDEN ECOSYSTEM

Some scientists think of ecosystems as multiple ecologies that interconnect and interact with each other. The original garden ecosystem consisted of several interlocking and nested ecologies. The most obvious is the "natural ecology" in which all of the animals, trees, rivers, and springs collectively created a lush natural environment perfectly suited to God's human handiwork. In addition, a "spiritual ecology" defined the relationship between God and his human creation. This spiritual ecology enabled the humans to flourish spiritually and enjoy God's fellowship free from any disturbance (Gen 3:8). There was a "social ecology" (even a "family ecology") represented in the bond between the first man and woman who became husband and wife (Gen 2:18-25). Built upon this social ecology was a "human ecology" that describes the individual functioning of the human (Gen 2:7, 15, 19-20). Each human being is a composite of connected and interacting systems and dimensions that operate harmoniously to sustain human life at an optimal level of existence as God intended (see figure 1). In the Genesis description of original humanity, we see physical, intellectual, spiritual, moral, and social aspects constituent of the whole person created in the image and likeness of God.

These subsets of the larger ecosystem of the Garden interacted with one another to create an interconnected whole. Within each ecological subset of the Garden ecosystem, we find specific ways in which the parts connect to the whole. The spiritual connection between God and humans manifested itself through interpersonal interactions that forged a significant spiritual bond between them. In the same way, the social connection between Adam and Eve forged a bond—"joined" them together—so strong that they became "one flesh" (Gen 2:24). Von Rad notices the strength of this original bond, writing, "Whence comes this love . . . stronger than the tie to one's own parents, whence this inner clinging to each other, this drive toward each other which does not rest until it again becomes the one flesh in the child?"[35]

---

[35]Gerhard Von Rad, *Genesis* (Philadelphia: Westminster Press, 1972), 85.

We can see in this original Garden ecology God created for the good of the first humans, all of the elements necessary for continued growth in perfection and righteousness.

THE PROBLEM OF SIN

However, into this perfect ecology, humanity introduced sin that disconnected us from our Creator and disrupted the entire ecological balance divinely created for us to enjoy. Ever since, humanity has searched to reestablish broken connections through religion, social contracts, political organizations, environmental policies, governmental edicts, and interpersonal relationships.

Each of the subsets of the larger ecosystem of the Garden of Eden (spiritual, natural, social, and human) felt the effect of human sin. All of the ecological connections and interactions that bonded God to humanity, humans to each other, and humans to nature and that made it possible for the human to live with personal integrity—*shalom*, "wholeness"—were disrupted and dislocated. All of the ecological harmony and synergy that existed prior to Genesis 3 fragmented into pixelated brokenness.

Human sin infected the connective harmony that permeated the pristine ecology God created. The entire web of connections and interactions that made God's original ecology hum along in perfect harmony was suddenly off-key and discordant. The scope of this disruption is breathtaking. None of the ecological subsets within the garden ecology could avoid the spread of this sinful infection. The word that perfectly describes these multiple disconnections is *alienation*. Alienation is a separation between things or persons previously connected. American theologian Paul Tillich, although hesitant to adopt the language of original sin, did embrace the concept of alienation, which has its roots in human sin that engendered human separation from God.[36] In his view, the result of this alienation is "estrangement" in all spheres of human existence.[37] Pannenberg critiques Tilllich's view of alienation and offers a Genesis-based elaboration that not only involves alienation from God but

---

[36]Paul Tillich, *Biblical Religion and the Search for Ultimate Reality* (Chicago: University of Chicago Press, 1955).
[37]Ibid., 55.

"entails a rupture between human beings and the natural world which God has created (Gen 3:17ff) and a rupture among human beings themselves (Gen 3:12 and 16)."[38]

Francis Schaeffer summarized the orthodox view of the fall, saying it "not only separated man from God, but also caused other deep divisions."[39] He went on to describe "psychological divisions," "sociological divisions," and the "ecological divisions" that separate humanity from nature and help explain the ecological crisis about which he wrote.[40]

There is some overlap between our analysis and Schaeffer's, but we see four ecological subsets, and thus four ecological alienations (disconnections): spiritual, social, human, and natural. Howard Snyder builds on Schaeffer's analysis and refers to these different ecological subsets as "the ecology of sin."[41] He writes, "All these divisions derive from sin, and all distort God's good purpose in creation. Therefore they all cry out for a gospel of reconciliation and healing."[42] Because of human sin, the spiritual connection between humanity and God, the social connection between humans, the natural connection between humanity and nature, and the internal wholeness of individual persons are broken beyond human repair. Human sin introduced alienation and disruption that could only be repaired through the finished work of Christ, who reconciled us to God and gave his people a ministry of reconciliation (2 Cor 5:18-20).

SUMMARY

As humans we live in a series of nested ecologies—all of which are connected in one form or another and therefore mutually influence each other to varying degrees. God created all of these ecological realities, including the ecology of the Garden of Eden—our first human home. Human sin disrupted the interconnected harmony of this divinely created, cosmic ecology and introduced the ecology of sin. In subsequent chapters,

---

[38]Wolfhart Pannenberg, *Anthropology in Theological Perspective* (Philadelphia: The Westminster Press, 1985), 284.

[39]Francis A. Schaeffer, *Pollution and the Death of Man: The Christian View of Ecology* (Wheaton, IL: Tyndale House, 1970), 67.

[40]Ibid.

[41]Howard A. Snyder, *Salvation Means Creation Healed: The Ecology of Sin and Grace* (Eugene, OR: Cascade Books, 2011), 65-80.

[42]Ibid., 68.

we will explore how God brings about the restoration of all of this ecological brokenness, alienation, and disconnection in Christ so that we can once again utilize the divine resources necessary for our full enjoyment of God's salvation.

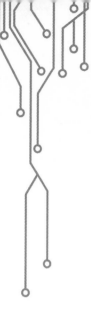

— TWO —

THE ECOLOGICAL MOTIF
IN SCRIPTURE

When we try to pick out anything by itself, we find
it hitched to everything else in the universe.

JOHN MUIR, *MY FIRST SUMMER IN THE SIERRA*

When American naturalist John Muir took his "thousand-mile walk" from his home in Indiana to Florida, which he wrote about in *A Thousand-Mile Walk to the Gulf*, he took his personal copy of the New Testament with him on his journey. While reading God's word outdoors, Muir developed a more holistic and protoecological appreciation of creation "that prepared the ground for a part of twentieth century ecological thinking in America."[1] Wendell Berry confirms the wisdom of Muir's approach when he writes, "I don't think it is enough appreciated how much an outdoor book the Bible is. . . . It is best read and understood outdoors, and the farther outdoors the better."[2] The biblical authors spent much time in the outdoors and from this vantage point brought their observations of its marvelous attributes into the words they wrote.

Gordon Zerbe affirms their vantage point when he writes, "From beginning to end, from first part to last part, the Bible is an ecological book."[3] He means "there is a remarkable commonality between the ecological and

---

[1]Jean-Daniel Collomb, "Christianity to Ecology: John Muir's Walk Through America," *Transtext(e)s Transcultures* 4, (2008): 100-113, http://transtexts.revues.org/254; doi:10.4000/transtexts.254.

[2]Wendell Berry, *Sex, Economy, Freedom, and Community: Eight Essays* (New York: Pantheon, 1993), 103.

[3]Gordon Zerbe, "Ecology According to the New Testament," *Direction* 21, no. 2 (1992): 15.

the biblical perspective."[4] He sees not only this commonality but also an ecological slant on the New Testament's concept of the kingdom of God as well as the concept of restorative wholeness to God's creation in the consummation. Ernst Conradie expands this observation to insist "it is important to read the whole Bible through ecological spectacles. This soon leads to the discovery that the Bible, from Genesis to Revelation, is 'filled to the brim' with ecological overtones."[5]

Figure 2. John Muir, circa 1912

Beyond simply noting an ecological emphasis in the Bible like Zerbe and Conradie, H. Paul Santmire uses the term "ecological motif" in contrast to what he calls the "spiritual motif" to describe two contrasting ways to view a Christian attitude toward nature.[6] While we use the term "ecological

---

[4]Ibid., 15.
[5]Ernst M. Conradie, *Christianity and Ecological Theology* (Stellenbosch: SUN Press, 2006), 69.
[6]H. Paul Santmire, *The Travail of Nature: The Ambiguous Ecological Promise of Christian Theology* (Philadelphia: Fortress, 1998), 9.

motif," we use it from within a biblical tradition rather than a theological one as Santmire does. We are using the term *motif* in its literary meaning as "a pattern that appears in a written text."[7] A biblical motif represents a repeated concept, such as the shepherd motif, the wilderness motif, or the Exodus motif, that frequently appears in Scripture. However, as Weston Fields reminds us, "In the field of biblical literature no one definition of 'motif' is agreed upon."[8] In addition, no normative hermeneutical controls exist for identifying a legitimate biblical motif. Most biblical writers treating a biblical motif simply assume everyone understands what constitutes a motif and proceed to develop one they have identified. Consequently, as Dennis Horton concludes, "Current usage of the term qualifies nearly anything as a motif."[9]

Though he does not apply any of his insights to Scripture, William Freedman offers us some guidance concerning literary motifs in general. Given the paucity of guidance for determining a biblical motif, we may legitimately use his criteria for our purposes.[10] Freedman distinguishes between the identification or "establishment" of a motif and its literary efficacy. We are interested in its identification and accept the fact that the use of the motif has spiritual efficacy for the faithful who read and study it. The first "indispensable factor" that Freedman notes for the identification of a motif is "the frequency with which it recurs."[11] The greater the frequency of occurrence of an image, theme, or verbal pattern, the more profound the impression upon the reader. The second identifying factor is "the avoidability and unlikelihood of the particular uses of a motif."[12] By this he means that "the more uncommon a reference is in a given context, the more likely it is to strike the reader" and grab attention.[13] What we mean by ecological motif certainly meets Freedman's first identifying factor of frequency and his

---

[7]Leland Ryken, James C. Wilhoit, and Tremper Longman III, eds., *Dictionary of Biblical Imagery* (Downers Grove, IL: IVP Academic, 1998), xv.

[8]Weston Fields, *Sodom and Gomorrah: History and Motif in Biblical Narrative* (Sheffield: Sheffield Academic Press, 2009), 19.

[9]Dennis Horton, *Death and Resurrection: The Shape and Function of a Literary Motif in the Book of Acts* (Eugene, OR: Wipf and Stock, 2009), 1.

[10]William Freedman, "The Literary Motif: A Definition and Evaluation," *NOVEL: A Forum on Fiction* 4, no. 2 (Winter 1971): 123-31.

[11]Ibid., 126.

[12]Ibid.

[13]Ibid.

second identifying factor of improbability; we might expect to find references to creation in general but not necessarily references to the ecological diversity and interconnectedness of creation.

We use *ecological* to mean that which pertains to the interrelationship and interconnection of all things in nature. The study of ecology directs our attention to what Howard Snyder refers to as "the most comprehensive conceptual frame we have for visualizing the complex interrelationships of factors that make up human life and the life of our planet."[14] An "ecological motif" then is a repeated pattern of reference in the biblical text that describes creation from an ecological vantage point that stresses nature's interconnections. Edwin Yamauchi reminds us that although the biblical writers did not possess a scientific understanding of ecology, they "were at least partially aware 'of the interrelationships of living things to one another and their surrounding environment.'"[15] Ellen Davis observes this ecological perspective early on in Genesis where "one can discern that the form of human life is fundamentally ecological" in the sense that "we are enmeshed in a harmonious web of relationships, infinitely complex in their intersections, that have in God their origin and their point of cohesion."[16]

ECOLOGICAL IMAGES OF GROWTH

Scripture abounds with images of growth that would have been easily understandable in its original context and to its original readers. Daniel Hillel says, "All the events described in the Bible took place in the distinctive ecological domains of the region that includes the Fertile Crescent."[17] This environmental reality may explain why biblical writers communicate the ecological motif in "implicit messages, expressed as allusions, associations, and connotations."[18] This ecological context would also seem to explain why the biblical writers borrow terms and references from this environment for

---

[14]Howard A. Snyder, *Salvation Means Creation Healed: The Ecology of Sin and Grace* (Eugene, OR: Cascade Books, 2011), xvi.

[15]Edwin M. Yamauchi, "Ancient Ecologies and the Biblical Perspective," *Journal of the American Scientific Affiliation* 32, no. 4 (1980): 193.

[16]Ellen F. Davis, *Scripture, Culture, and Agriculture: An Agrarian Reading of the Bible* (New York: Cambridge University Press, 2009), 56-57.

[17]Daniel Hillel, *The Natural History of the Bible: An Environmental Exploration of the Hebrew Scriptures* (New York: Columbia University Press, 2006), 13.

[18]Ibid., 21.

their own purposes within their distinctive narrative traditions. One of those purposes is to illustrate to Israel the ecological wholeness of land as God's creation. Richard Bauckham argues that we should not "lose sight of the inter-connections and interdependence . . . which modern ecology reveals in scientific detail" in passages such as Job 38–41.[19] In these chapters, natural elements intertwine with the animal kingdom in a symphony of natural ecology.

Psalm 104 is another example of how the biblical writers paint a "picture of an ecological creation."[20] It depicts interdependency in God's creation and values each segment of the ecosystem. No living creature can exist apart from its connection with other living creatures. The psalm depicts the diversity of living creatures and assigns a habitat for each. In Psalm 104:16-23, birds build their nests in trees, the wild goats live in the mountains, beasts live in the forests, etc. In Psalm 104:14-15, plants and other vegetation provide food for animals and sustains human life. And in Psalm 104:16-17, the trees of Lebanon absorb the rain and provide a place for the birds to nest. The psalmist looks down at the grass and up at the mountains, and between these two vantage points he sees a divinely orchestrated ecosystem where "everything is connected to everything else."[21] This is exactly what John Goldingay recognizes when he sees "the ecology of creation" expressed as "the interrelationships of creation" manifesting an "interwoven ecology," likened to "a magnificent quilt in which every thread contributes to a whole, woven by a supremely skilled craftworker."[22]

Paul famously appeals to this imagery in 1 Corinthians 3:6: "I planted, Apollos watered, but God was causing the growth." God is the ultimate source of all living and growing things in his creation, but in this passage, God utilized the spiritual farming efforts of Paul and Apollos to achieve growth. Growth is a byproduct of an intricate series of interlocking events, conditions, and exchanges between and among living and nonliving entities. When we zoom out and take a Google Maps view of growth in God's creation, we see the interconnected and interactive elements that conspire

---

[19]Richard Bauckham, *The Bible and Ecology: Rediscovering the Community of Creation* (Waco, TX: Baylor University Press, 2010), 146.
[20]Ibid., 70.
[21]Ervin Laszlo, *The Interconnected Universe* (Singapore: World Scientific), 9.
[22]John Goldingay, *Psalms*, vol. 3, *Psalms 90–150* (Grand Rapids: Baker Academic, 2008), 190-91.

to instigate, encourage, and reward growth. We are on solid ground when we say all growth in God's creation is ecological growth, for nothing grows alone and detached from other living and growing things and persons. The portrait Paul paints in 1 Corinthians 3:6 is one that illustrates the ecological interplay between planting, watering, and growth. While Paul was addressing the issue of spiritual growth in the Corinthian community, he borrowed terms and processes most often associated with agriculture. Agricultural references may explicitly identify ecological aspects of growth or assume them to be implicit. Most of the references we see in Scripture take an implicit stance since most of the biblical authors took these matters to be common knowledge. As Hillel observes, "Although the natural environment seldom serves as a subject in its own right, it is reflected indirectly as the backdrop to described events."[23] Scripture contains references to several particular aspects of the created order, each of which demonstrates ecological growth.[24]

Trees: Psalm 1:3, Jeremiah 17:7-8, and Ezekiel 31:4-5. Psalm 1:1-3 draws a comparison between the ecology of trees and the righteous person—who like the tree is "planted by streams of water" and "yields its fruit in its season" (Ps 1:3). Here we find a prepositional phrase that denotes a defined ecology, "by streams of water." Trees do not grow alone; they grow as they connect to and interact with a greater ecology that provides part of the nourishment and nutrients needed to sustain life and produce growth. The tree also contributes to the ecology in which it lives by adding nutrients to the soil and atmosphere and by hosting birds who build their nests in its limbs.[25] All of this the psalmist understands, not only about the ecology of trees but the ecology of the righteous person who does not flourish alone but as he or she is planted within a defined ecology. Righteous people avoid the detrimental social ecology described in Psalm 1:1—"counsel of the wicked," "path of sinners," "the seat of scoffers"—and places themselves within the beneficial ecology of the law of God and of God's people who follow that law.

---

[23]Ibid., 247.

[24]See here Ryken, Wilhoit, and Longman, *Dictionary of Biblical Imagery*.

[25]Robert Leo Smith, *Ecology and Field Biology* (Menlo Park, IL: Addison Wesley Longman, 1996), 263-88.

All three texts refer to the growth of trees in a defined ecosystem expressed by four prepositional phrases: "by streams of water," "by the water," "by a stream," and "trees of the field." All three texts highlight the power of the ecosystem to induce flourishing ("its leaf does not wither," "its leaves will be green," and "its height was loftier . . . its boughs became many and its branches long"). In two of the three texts, the biblical writers draw parallels between the righteous and the flourishing growth of trees in a defined ecology. The flourishing trees and the righteous give evidence of health and vitality through observable indicators that confirm growth while also confirming the viability of the natural or spiritual ecosystem that produced them.

Vineyards: Psalm 80:8-16, Jeremiah 2:21, Isaiah 5:1-7, Ezekiel 17:7-10, Amos 9:13-15, and John 15:1-11. Within the larger cosmic ecology of the universe, many discreet ecological communities exist. Although often studied as distinct ecological habitats, such as lakes and forests, they are not independent of the wider ecosystem of which they are a part. The vineyard fits the definition of a discreet ecological community or mini-ecosystem.[26] A descriptive account of the vineyard as a mini-ecosystem with defined boundaries appears in Isaiah 5:1-7, in the commonly identified "Song of the Vineyard." Martin Klingbeil highlights this text because it "presents an important aspect of creation in demonstrating the interconnection of God's creation."[27] Interconnection is the hallmark of an ecological understanding of creation that recognizes and values the mutual resources these interconnections provide for the growth of all living things and persons.[28]

This text also has the advantage of declaring, unambiguously, "The vineyard of the LORD of hosts is the house of Israel" (Is 5:7). It thereby makes a direct correlation between vineyard imagery and Israel. In this text, the owner of the vineyard not only planted hedges around the vines but also took the added precaution of building a wall around it (Is 5:5). Here we see most clearly that a vineyard represented a special type of ecosystem with defined boundaries that demarcated it but did not separate it from the surrounding ecological environment. The defined boundaries of the vineyard

---

[26]G. T. McGourty, J. Ohmart, and D. Chaney, *Organic Winegrowing Manual* (Richmond: The University of California Division of Agriculture and Natural Resources, 2011).

[27]Martin G. Klingbeil, "Creation in the Prophetic Literature of the Old Testament: An Intertextual Approach," *Journal of the Adventist Theological Society* 12, nos. 1-2 (2009): 38.

[28]Bauckham, *The Bible and Ecology*, 146-51.

ecology are evident by the physical boundary marker of the wall or hedge that surrounded the vineyard of a particular owner. The story of Naboth's vineyard in 1 Kings 21 highlights the family and covenant boundary markers that kept the vineyard in patrilineal succession over the generations.[29] The combination of the physical, covenantal, and familial boundary markers of the vineyard serve to emphasize the fact that a vineyard was a mini-ecosystem and that Israel manifested similar ecological characteristics from a sociological and spiritual perspective.

What arrests our attention in Isaiah's description of Israel as the "vineyard of the LORD of hosts" is that although the Lord "expected *it* to produce *good* grapes, . . . it produced *only* worthless ones" (Is 5:7, 2). One would expect—given the due diligence of the Lord expressed in five verbs of vineyard care (dug, removed, planted, built, hewed) and given that the Lord located the vineyard on a "fertile hill" and built a double fence (hedge and wall)—that the choicest of fruit would result with an abundant harvest (Is 5:1, 5). We would expect the defined mini-ecosystem of the vineyard with the organic interconnections and mutual interactions of shared resources and the care motivated by love would result in prolific growth (Is 5:1). When we do not see what we expect, we are forced to identify internal or external threats to the ecosystem. Isaiah tells us that the threat to the vineyard ecology was internal—derived from Israel's "bloodshed" and oppression of the marginalized that produced a cry (scream) of distress (Is 5:7).[30]

Both the Old Testament (*kerem*) and the New Testament (*ampelōn*) use the vineyard to symbolize the nation of Israel (Is 5:1-7; Jer 2:21; 6:9), the kingdom of God (Mt 20:1-11; 21:33-43), or the followers of Jesus (Jn 15). There are also references to literal vineyards, which were commonplace in ancient Israel (Jer 31:5; Amos 9:14; Zeph 1:13).

Psalm 80:8-16 is another passage typical of a metaphorical use of vineyard imagery to refer to Israel. The vineyard metaphor in Psalm 80:8-11 expresses what happened to Israel before the internal threats identified by Isaiah began to disrupt God's vineyard ecology. In this portion of the psalm, Asaph

---

[29]Carey Ellen Walsh, *The Fruit of the Vine: Viticulture in Ancient Israel* (Winona Lake, IN: Eisenbrauns, 2000).

[30]See J. Alec Motyer, *The Prophecy of Isaiah: An Introduction and Commentary* (Downers Grove, IL: InterVarsity Press, 1993), 68-69.

reminds God's people that God took a shoot from a vine in Egypt and planted it in the land of promise (Ps 80:8). The vine took root and became a vineyard that "filled the land" to such an extent that "the mountains were covered with its shadow" and "it was sending out its branches to the sea and its shoots to the River" (Ps 80:9-11). Goldingay likens the imagery to "a giant plant like Jack's beanstalk" from the children's story.[31] The hyperbolic language of the psalm conveys the impression of extensive growth within an ecosystem, with its boundaries marked by the Mediterranean Sea and the Euphrates River, comprising the biblical "narrative from Joshua to David."[32]

Both of these Old Testament texts provide an ecological perspective on growth or the lack thereof. As we noted from Psalm 80, Asaph rejoiced over the realized growth of Israel planted as God's vineyard in the land of promise. When properly functioning, ecological growth derived from organic interconnections and the mutual exchange of resources tended by a loving vintner is extensive and impressive. The covenant connection with the Lord, when intact and functioning, provides all the spiritual resources the Lord's vineyard needs for growth. Isaiah 5 says that eventually all of this growth stymies and reverts to "briars and thorns" because "the vineyard of the LORD of hosts" turned her back on her only source of spiritual nourishment (Is 5:6-7).

Of course, in the New Testament, the most significant passage that depicts the organic connection between the Lord and his people is the extended description in John 15:1-11. We can only approach the general orientation and flow of the passage as it pertains to the subject at hand. Few biblical scholars have allowed actual first-century viticulture practices to inform this vivid picture Jesus offers of his future relationship to his disciples (and those who will follow them). The relationship Jesus describes appertains after he goes to his Father and comes to them again through the ministry of the Holy Spirit (Jn 14:4, 12, 16, 18, 25-26). Despite the physical separation between Jesus and his disciples, he insists on using a rich organic and ecological image of their relationship to stress its intimacy and connectedness.

The vine and vineyard imagery depicts a reciprocal relationship (with phrases such as "Abide in Me, and I in you," "He who abides in Me and I in

---

[31]John Goldingay, *Psalms*, vol. 2, *Psalms 42–89* (Grand Rapids: Baker Academic, 2007), 539.
[32]Ibid., 539.

him," and "If you abide in Me, and My words abide in you") indicative of mechanisms of growth in a natural ecology (Jn 15:4-7). Worster says this is a basic feature of any ecosystem: "The health of an ecosystem also required a condition of mutualism and cooperation among the many organisms inhabiting an area."[33] This description of the essential quality of ecological functioning gave rise to the study of ecology as "a relational discipline speaking a relational language."[34] The relational quality of ecosystem functioning that operated in the vineyard ecology provided Jesus the perfect illustration of his relationship to his followers. As we will discuss in a subsequent chapter, Jesus frequently turned to the natural world to illustrate spiritual principles operating in the kingdom and in his relationship with his followers. Most often, his borrowed analogies and metaphors reflected the ecological realities of growth in creation that stress organic interconnections and reciprocal interactions of shared nutrients producing mature, mutual growth.

Flowers: Psalm 103:15, Hosea 14:5, and Matthew 6:28. Biblical writers referred to flowers to illustrate beauty (Mt 6:28), love (Song 2:2), transient human mortality (Ps 103:15-16), and spiritual restoration (Is 35:1). While all of these uses, and many others, provide valuable insight, we want to focus our attention exclusively on the use of flower imagery to convey the concept of ecological growth. Sometimes these uses overlap, as in Hosea 14:5-7: "I will be like the dew to Israel; He will blossom like the lily, And he will take root like *the cedars of* Lebanon. His shoots will sprout, . . . And they will blossom like the vine."[35] The context of Hosea makes it clear that the prophet envisions a future time when God will revisit his covenant blessings upon Israel. One of the illustrations of this spiritual renewal is the blossoming and growing lily. Another example of a mixed use of the flower imagery is in Psalm 103:15 (NIV): "The life of mortals is like grass, they flourish like a flower of the field." While this passage certainly illustrates human transience

---

[33]Donald Worster, *Nature's Economy: A History of Ecological Ideas*, 2nd ed. (Cambridge: Cambridge University Press, 1994), 366.

[34]Ibid., 318.

[35]The original Hebrew text does not have any words between "like" and "Lebanon." Notice that the NASB used here italicizes the words "the cedars of" indicating that the translators are supplying these words, not translating them. Some interpreters supply "lilies of" as a possible translation of what the author originally intended. It makes more sense that Hosea would maintain the metaphor rather than mixing metaphors and thereby possibly confusing the reader.

("the life of mortals is like grass"), it also illustrates human flourishing and growth ("they flourish like a flower of the field"). The many references in both testaments to flowers of the field identify various types of flowers "growing in valleys, among brambles or in pasture land."[36] The prepositional phrase "of the field" (*śādeh*) highlights the ecosystem within which the flowers grow. John Hartley proposes that "this masculine noun broadly designates . . . a definite portion of ground."[37] This description of the term "field," along with the force of the prepositional construct, parallels our proposal: we think of the term "field" as a defined ecology within which various plants, seeds, flowers, and trees grow.

The phrase "flower(s) of the field" is found in the Old Testament but finds a more specific parallel on the lips of Jesus. In Matthew 6:28, we find a movement from lesser to greater in God's care for flowers and his creatures, where Jesus admonishes his disciples to "observe how the lilies of the field grow; they do not toil nor do they spin." He arrests our attention with the phrase "observe how the lilies of the field grow." The word Jesus used for "observe" (*katamathete*) in its intensive form (*kata* + *manthanō*) means "to consider, to observe, to think about."[38] Jesus called upon his disciples (and us) to focus and give attention to "how" lilies grow. The natural growth of lilies should have taught the disciples something about anxiety, while simultaneously teaching them something about the growth process. The third element is the prepositional phrase "of the field." That phrase, like its Old Testament equivalent, highlights the context or ecosystem within which the growth of lilies occurs. The entire phrase reminds us that lilies do not grow in isolation but within a rich and diverse ecology comprising soil, water, sun, other organisms, and insects, such as the honeybee. Granted, Jesus does not elaborate all of this ecological detail; he assumes that his disciples know this or will come to know it as they scrutinize the growth of field lilies.

Seed(s): Isaiah 30:23, Ezekiel 17:5-6, Matthew 13:31-32, and Mark 4:8, 26-29. As with most of the Hebrew prophets, Isaiah not only foresees a day of judgment on Israel for its covenant infidelity but also expects a day of

---

[36]Ryken, Wilhoit, and Longman, *Dictionary of Biblical Imagery*, 294.

[37]R. Laird Harris, Gleason L. Archer, and Bruce K. Waltke, *Theological Wordbook of the Old Testament* (Chicago: Moody Press, 1980), 2:871.

[38]Johannes P. Louw and Eugene A. Nida, *Greek-English Lexicon of the New Testament: Based on Semantic Domains*, 2nd ed. (New York: United Bible Societies, 1989), 354.

blessing and renewal. Isaiah 30:23 describes the latter: and in that day "Then [God] will give *you* rain for the seed which you will sow in the ground . . . and it will be rich and plenteous." There are two ecological references in the rain and the soil, which will enable the "rich and plenteous" harvest God will give Israel. God uses the natural elements and patterns of growth he established in his creation to bring about this fruitfulness. Both fruitfulness and unfruitfulness are part of the illustrations in Ezekiel 17, Matthew 13, and Mark 4. We explore the Gospel passages in greater depth in chapter three.

Ezekiel 17:5-6 utilizes a parable of two eagles to illustrate what God (the first eagle) did for Israel as his "seed" and how Israel rejected his provision for a lesser eagle, Babylon (Ezek 17:12). The first eagle "took some of the seed of the land and planted it in fertile soil. He placed *it* beside abundant waters. . . . Then it sprouted and became a low, spreading vine . . . and yielded shoots and sent out branches" (Ezek 17:5-6). Here again we have ecological references identified by land, fertile soil, and abundant waters, which place the seed within an ecosystem that provides all the nourishment and nutrients necessary for growth. In such a lavish ecosystem, we are not surprised by the resultant growth of the seed into a "spreading vine" that produces "shoots and [sends] out branches." The description suggests that seed grew sequentially ("sprouted," "became a vine," and "yielded shoots") and extensively ("sent out branches"), but it also reproduced itself through the mutual exchange of resources—as all functioning ecologies do.

While commentaries on Ezekiel explain the details of this riddle, we direct your attention to the description of Israel's growth when properly connected to God ("turned toward him," Ezek 17:6). Ezekiel presents Israel's spiritual health and vitality using ecological images of growth. These images convey important principles and patterns of spiritual growth we mustn't overlook. When we consider the purposely chosen image of a rich and diverse ecosystem within which Israel flourished in her relationship with God, we think about spiritual growth within God's perspective—not one created out of our own deliberations or theological traditions. The spiritual ecosystem that God provided Israel consisted of fertile soil and abundant waters. The spiritual growth of Israel mirrored the natural growth of seed properly planted, watered, and tended. Such spiritual growth reflects natural patterns of growth that follow a movement from immature seed to mature vine and branches. Here, in a context of a contrived riddle, the growth is idealistic; we know from other passages (Ps 30; 40) that such growth rarely comes without the presence of what Walter Brueggemann calls the "wrenching transitions" or "points of disjunction."[39] We see these transitions and disjunctions sprinkled throughout the Old Testament but featured most prominently in psalms of "disorientation" (Ps 13; 22; 86).[40]

In this Ezekiel passage, the point of disjunction comes through Israel's rejection of God's provision and the resultant exile in Babylon. This change of ecosystem poses a question uttered twice in the passage by the Lord: "Will it thrive?" (Ezek 17:9-10). Will the nation of Israel thrive when it has "sent out it branches toward him (the second eagle) from the beds where it was planted?" (Ezek 17:7). When Israel turned toward Babylon, it simultaneously turned away from the Lord, sparking the rhetorical question: "Will it thrive?" The obvious answer from the historical context is not only will Israel *not* thrive but "will be scattered to every wind" (Ezek 17:20-21). We can only expect thriving seeds and plants when they are properly situated within the appropriate environment where all the requisite nutrients stimulate growth.

Mixed ecological domains: Psalm 65:9-13, Joel 2:21-22, and Amos 9:13-15. Several texts express the ecological fecundity God provided for Israel in the land of promise. This rich fertility embraces all ecological domains: "the pastures," "the hills," "the meadows," "the valleys," and even "the wilderness." There is not only growth across this verdant ecological landscape but abundant growth ("drip with fatness," "covered with grain," and "yielded in full"). The two texts from Joel and Amos describe what God will do for Israel through the land upon their return from exile and disgrace. Not only will God restore them to himself spiritually and to the covenant but he will also restore their land to a greater degree than before the exile. It is clear from these texts that Scripture does not support the Platonic division between natural and spiritual. God employs the natural, created realm to accomplish redemptive purposes for his people, and he displays his redemption through the natural order.

---

[39]Walter Brueggemann, *Hope Within History* (Atlanta: John Knox Press, 1987), 8-9.
[40]Walter Brueggemann, *Praying the Psalms: Engaging Scripture and the Life of the Spirit* (Eugene, OR: Wipf and Stock, 2007), 11.

SUMMARY

It is evident that one can make a case for recognizing the ecological motif throughout both Old and New Testaments underlining the fecundity, diversity, and dynamic growth potential of living organisms in a defined ecosystem. Many biblical writers use this ecological reality to illustrate how God's people grow spiritually in their relationship to God and to one another. The ecological motif makes clear that God designed his creation to thrive and flourish within the verdant confines of a natural ecosystem intended to maximize growth and support reproductive abilities. None of the various species of plants, flowers, trees, and seeds grow in isolation, sequestered from other living organisms. Instead, their interconnectedness and mutualistic patterns of interrelationships create reciprocal biofeedback loops for sharing nutrients and resources—all for the benefit of the whole. The Bible teaches that God's people have a relationship to God and to one another built upon a deep spiritual connection of faith and fidelity (*ḥesed*). Our connection to Jesus, which he likened to vine and branches growing in a vineyard, makes possible our connections to other Christians. These spiritual connections serve as interactive conduits for the exchange of spiritual nutrients and resources that lead to mutual growth and maturity. In the next chapter, we will explore what Jesus understood to be the ecological character of growth in creation and how he used it to illustrate to his disciples how growth occurs in the kingdom of God.

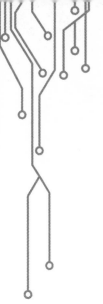

THREE

THE ECOLOGICAL POWER
OF PARABLES

*The seed grows from stalk to ear, and from
ear to ripened corn—the naming of each stage of the
process describes the unceasing process of growth.*

<small>JOACHIM JEREMIAS, *THE PARABLES OF JESUS*</small>

The most distinctive feature of the public teaching ministry of Jesus was his use of parables to instruct his followers and challenge the thinking of the multitudes who followed him. While parable scholars debate whether Jesus' use of parables was unique or if it fit within an established Jewish tradition, German scholar Joachim Jeremias was unequivocal: "Jesus' parables are something entirely new. In all the rabbinic literature, not one single parable has come down to us from the period before Jesus; only two similes from Rabbi Hillel who jokingly compared the body with a statue, and a soul with a guest."[1] James Breech conducted an exhaustive study of stories in existence from the time of Alexander the Great (300 BC) to Constantine (300 AD) and reached a similar conclusion when he wrote, "Jesus' parables are utterly dissimilar from any other stories known in Hellenistic and Greco-Roman antiquity, including Rabbinic parables."[2]

Simon Kistemaker writes, "He was knowledgeable in farming, sowing seed, detecting weeds, and reaping a harvest. He was at home in the

---

[1]Joachim Jeremias, *The Parables of Jesus* (New York: Charles Scribner and Sons, 1972), 19.
[2]James Breech, *Jesus and Postmodernism* (Philadelphia: Fortress, 1989), 64.

vineyard, knew the times of reaping fruit from vine and fig tree."[3] Richard Bauckham illustrates the knowledge Jesus possessed about the natural and agrarian world he inhabited, noting that Jesus referred to "bramble, fig tree, herbs (mint, dill, cumin, rue, and others), mulberry, mustard plant, reed, thorn, vine, weed, wheat, and wild flowers."[4] He was also acutely aware of the weather and the different seasons (Lk 12:54; Mk 11:13; Jn 4:35). Jesus did not attain this knowledge by being an effective farmer. As cocreator of all things, Jesus knew his masterpiece intimately and perfectly. He was able to do what no other person could to help his disciples understand the processes of growth.

The parables of Jesus help us see creation as Jesus did, and through his teachings we are able to understand the wholeness of life and salvation—or, as Howard Snyder puts it, to see the "ecology of salvation."[5] Robert Alter and Frank Kermode provide a schematic perspective on the gift of seeing from the story of Balaam and Barak in Numbers 22–24. Alter and Kermode write that the most prominent Hebrew verb in the familiar story is "to see" (*r'h*). The pagan king Balak sees the way in which God has blessed Israel by its military exploits and extensive land holdings (Num 22:2, 5; 23:9). Balaam self-identifies as a seer who sees the future (Num 24:3-4). However, Balaam cannot see the presence of an armed angel blocking the path of his donkey until "the LORD opened the eyes of Balaam, and he saw the angel of the LORD standing in the way with his drawn sword in his hand" (Num 22:31). With this humorous story, Alter makes the point that despite all of his proud assertions about his prowess as one who can see the future, "Balaam can see only what God reveals."[6] In our natural human condition, we cannot accurately see what God wants us to see until he reveals it to us. We need God's truth revealed to us so that we can see what God sees. Jesus opens our eyes to see the processes of spiritual growth perfectly illustrated in the growth of flowers, seeds, and trees.

---

[3]Simon J. Kistemaker, *The Parables of Jesus* (Grand Rapids: Baker Academic, 1980), xvii.

[4]Richard Bauckham, "Reading the Synoptic Gospels Ecologically," in *Ecological Hermeneutics: Biblical, Historical and Theological Perspectives*, ed. David G. Horrell, Cherryl Hunt, Christopher Southgate, and Francesca Stavrakopoulou (London: T&T Clark, 2010), 73.

[5]Howard A. Snyder, *Salvation Means Creation Healed: The Ecology of Sin and Grace* (Eugene, OR: Cascade Books, 2011), 78.

[6]Robert Alter and Frank Kermode, *The Literary Guide to the Bible* (Cambridge, MA: Harvard University Press, 1987), 86.

PARABLES OF GROWTH

We see most clearly the relationship between the process of natural growth and spiritual growth applied to the kingdom in the so-called "parables of growth" or "nature parables." G. R. Beasley-Murray counts among his parables of growth (1) the mustard seed and leaven, (2) the seed growing secretly, (3) the sower, (4) the wheat and tares, and (5) the dragnet.[7] Although he correctly includes these because of the similarities they bear to each other, we delimit our attention to only those parables of growth that connect a natural element of growth with growth in the kingdom of God. We are most interested in these specific parables because they speak to the issue of kingdom growth through the lens of some unique feature of natural growth in creation. Referring to these parables of growth, Richard Bauckham notes, "So in these parables the comparison is between God-given growth in creation and the God-given growth of the Kingdom of God, or, we might say, between the divine work of creation and the divine work of salvation and renewal."[8]

We will not enter into the scholarly debate over the kingdom significance of the growth element in the parables. We take our cue from the synthesizing work of N. A. Dahl, who directly addresses the concept of growth in the parables of growth. Dahl rejects the idea "that the point of the parables is this idea of growth." Rather, he argues, "it is presupposed as a matter of course."[9] He also insisted that the ancients represented in both the Old and New Testaments, although not possessing a contemporary scientific concept of growth, nevertheless understood and employed natural growth processes observed from experience. In particular, he writes, "To the growth which God in accordance with his own established order gives in the sphere of organic life, corresponds the series of events by which God in accordance with his plan of salvation leads history towards the end of the world and the beginning of the new aeon."[10] Dahl's point here is that there is a correspondence between growth in the "sphere of organic life" and growth "in accordance with his plan of salvation" taught in the

---

[7]G. R. Beasley-Murray, *Jesus and the Kingdom of God* (Grand Rapids: Eerdmans, 1986), 194-201.

[8]Richard Bauckham, "Reading the Synoptic Gospels Ecologically," in *Ecological Hermeneutics: Biblical, Historical and Theological Perspectives*, ed. David G. Horrell et al. (London: T&T Clark, 2010), 76.

[9]N. A. Dahl, "The Parables of Growth," *Studia Theologica* 5, no. 2 (1951): 132-66, 146.

[10]Ibid., 146.

parables of Jesus.[11] Other parable scholars share this view of the role of growth in the parables of growth.

British parable scholar A. M. Hunter helps us understand the relationship between nature and kingdom. He says a parable is "a comparison drawn from nature or daily life and designed to illuminate some spiritual truth, on the assumption that what is valid in one sphere is valid also in the other."[12] Jesus used his expert knowledge of nature "to illuminate some spiritual truth" about the kingdom that his disciples needed to learn.[13] He goes on to suggest that Jesus taught this way based upon "the assumption that what is valid in one sphere is valid also in the other."[14] Jesus emphasized the process of growth as the main point of comparison in these parables. That is, according to Hunter, Jesus assumed that the processes of growth in nature ("one sphere") were valid also in the kingdom ("the other").

C. H. Dodd argues something similar when he writes about the source of the stories Jesus told:

> There is a reason for this realism of the parables of Jesus. It arises from the conviction that there is no mere analogy, but an inward affinity, between the natural order and the spiritual order; or as we might put it in the language of the parables themselves, the Kingdom of God is intrinsically *like* the process of nature and of the daily life of men.[15]

By this Dodd means that "nature and super-nature are one order," so by "affinity" he means *identical* since both the natural and the spiritual are "one order."[16] Dodd refers to this affinity between natural and spiritual realms as "the major premise of all the parables."[17]

One simple way of confirming Hunter, Dodd, and Dahl's common assertion is to note an often-overlooked New Testament reality. New Testament writers predominantly used a Greek word for growth (*auxanō*) whether they were referring to growth in nature (Mk 4:8), human beings (Lk 1:80; 2:40), the kingdom (Mt 13:31-32; Mk 4:8; Lk 13:19), the church (Eph 2:21;

---

[11]Ibid.
[12]Archibald M. Hunter, *Interpreting the Parables* (Philadelphia: The Westminster Press, 1960), 8.
[13]Ibid.
[14]Ibid.
[15]C. H. Dodd, *The Parables of the Kingdom* (London: Nisbet, 1946), 20.
[16]Ibid., 10.
[17]Ibid.

4:15; Col 2:19), or the Christian (1 Pet 2:2; 2 Pet 3:18). Leroy asserts that "behind [the New Testament] use of the verb lies the image of growth in the realm of creation."[18] We see this most clearly in 1 Peter 2:2 where Peter draws a straight line between how "the newborn infant grows physically on the basis of milk" and how "the believer grows spiritually on the basis of the word."[19] As far as Scripture is concerned, growth is growth, whether we see that growth in the natural ecology of creation or the spiritual ecology of the church and kingdom. God did not establish two separate laws of growth— one governing flowers and trees and another governing the kingdom and the church. Growth in nature and growth in the kingdom, the church, and the Christian partake of essentially/virtually identical patterns that require ecological connections and reciprocal interactions expressed as nutrient exchanges. In nature, the connections and exchanges are organic, while in the kingdom and church, they are spiritual. In nature, the exchanges involve physical nutrients (carbon, nitrogen, oxygen, etc.); in the church and kingdom they involve spiritual nutrients (milk and meat of the Word) exchanged through interactions with one another.[20]

THE PARABLE OF THE SEED GROWING ECOLOGICALLY: MARK 4:26-29

We find one of the most familiar parables of Jesus in Mark 4:26-29, where he illustrates the growth of the kingdom by reference to the growth of seeds. Mark situates this particular parable between the parable of the seeds and soil and the parable of the mustard seed (Mk 4:1-12, 30-34). All three of the parables have in common the elements of seeds, soil, growth, and kingdom. Although the interpreter, as David Garland says, has exegetical options as to the identity of "the key elements of the parable," for our purposes, we focus attention only on the growth of the seed.[21]

The parable opens with an actively engaged farmer "who casts seed upon the soil" and then goes about his normal farming routines ("goes to bed at

---

[18]H. Leroy, "*Auxanō, auxō*," in *Exegetical Dictionary of the New Testament*, ed. Horst Balz and Gerhard Schneider (Grand Rapids: Eerdmans, 1999), 1:178.

[19]Ibid.

[20]We explore reciprocal interactions (*allēlōn*) that involve the mutual exchange of spiritual nutrients between believers in part 3.

[21]David E. Garland, *Mark*, NIV Application Commentary (Grand Rapids: Zondervan, 1996), 177.

night and gets up by day," Mk 4:26-27). While he carries out his daily farming duties, "the seed sprouts and grows" but how this growth occurs "he himself does not know" (Mk 4:27). This is probably the most startling element of the parable. The original audience, which is comprised of those who regularly engaged in the activities described in the parable, would no doubt register shock at the ignorance of the farmer in the parable. As an agricultural expert, one would expect the planter and harvester of the seed to know something about how it grows. Jesus adds this surprise element to emphasize the same thing Paul did in 1 Corinthians 3:6; although Paul planted and Apollos watered, God caused the growth of the Corinthian church. In the same way, the farmer could plant and harvest, but he could not make seeds grow. His ignorance of the process indicates as much.

The parable teaches us that we cannot make anything grow on our own. We can only set the ecological conditions for growth. The farmer fulfills his or her role and God fulfills his. The farmer "casts seed upon the ground," and he "puts in the sickle" when the harvest is ready. The soil does its work silently while the farmer goes about his daily activities ("he goes to bed at night and gets up by day," Mk 4:27). The farmer creates an ecology conducive to growth and then lets the process of growth unfold by itself (*automatē*, Mk 4:28). This word, *automatē*, is key in the parable, though often overlooked. The Greek word means "without any cause, without something to cause it, by itself."[22] This is the same word used in Acts 12:10 to describe how Peter's prison door opened "by itself." Each seed contains within itself the required mechanisms that will establish the seed in the soil and permit it to receive further nutrients to sustain the process of growth to maturity.

The process of seed growth follows an invariant developmental trajectory: "first the blade, then the head, then the mature grain in the head" governed by innate patterns (*automatē*) combined with environmental factors supplied by the defined ecology of the planted field (Mk 4:28). The textual sequence, "first . . . then . . . then," identifies the invariant sequence of growth from one stage to another—sequentially, without skips—until it reaches full maturity. Just like the images of ecological growth we saw in the

---

[22]"*Automatos*," in Johannes P. Louw and Eugene A. Nida, *Greek-English Lexicon of the New Testament: Based on Semantic Domains*, 2nd ed. (New York: United Bible Societies, 1989), 780.

Old Testament and so many of the parables of Jesus, here too we find the culmination of the growth process in a harvest, indicating the efficacy of the ecology governed by God's laws to produce abundantly.

Perhaps the point of the parable is to remind us to have faith in the processes of growth that God has established in nature and the kingdom. There is nothing contradictory in the parable between the active role of the farmer in certain tasks (sowing and reaping) and the active role of God in producing growth. The parable highlights what Peter Jones calls "a dramatic tension between the sleeping man and the sprouting seed."[23] Ultimately, however, the seed grows because of the combined efforts of the farmer and God. If the farmer did not perform his duties of preparing the soil, tilling the soil, and sowing the seed (which God does not perform) there would be no seed in the ground to grow by itself (*automatē*). If God did not create a world in which seeds grow *automatē* when planted in the proper environment, there would have been no harvest for the farmer to reap. Perhaps we should say that ecological growth leads to eschatological abundance. There will be no future harvest if the farmer does not plant the seed properly and cultivate it carefully in an environment conducive to growth. The seed grows ecologically in accordance with God's design for growth, which operates in the natural field and the spiritual field (Mt 13:24; 1 Cor 3:9).

THE PARABLE OF THE SOWER AND SOIL ECOLOGY: MATTHEW 13:3-9, MARK 4:1-9, AND LUKE 8:4-8

The importance of a proper ecology for seeds receives special attention in the parable of the sower found in all three Synoptic Gospels. Not just any soil environment will sustain the growth of a seed. The prepositions in these verses that link the relationship of the seed to the soil are crucial; some of the scattered seed "fell beside [*para*] the road," while some "fell on [*epi*] rocky *soil*," some "fell among [*en*] the thorns," and some "fell into [*eis*] the good soil" (Lk 8:5-8). The first three prepositions indicate a proximate relationship of seed to soil, but only the fourth preposition (*eis*) describes an intimate relationship between seed and soil. The first three prepositions indicate

---

[23]Peter R. Jones, "The Seed Parables in Mark," *Review and Expositor* 75 (1978): 526.

that the seed stood very little chance of germinating and growing simply because it did not find a hospitable home *in* the soil; it was rather *beside, on,* or *among* its environment. Greek prepositions originally conveyed spatial relationships involving motion to (*eis,* "into"), motion from (*apo,* "away from"), and rest (*en,* "in" and *epi,* "upon" and *para,* "beside").[24] The first three prepositions in Luke's rendition of the parable are prepositions of rest, while the final one is a preposition of direction—*to* or *into.*

The ideal soil ecology for the scattered seed is obviously the final one ("good soil") because it was *in* this soil that the seed "grew up, and produced a crop a hundred times as great" (Lk 8:8). The other soil environments did not allow the seed to sink into the soil, germinate, and grow. As with most parables, the storyteller avoids extraneous details, but we can assume that the original audience filled in the details as they heard the story. They would have filled in details about soil preparation, cultivation, nutrients, and irrigation. Daniel Hillel informs us that sustaining crops in Israel requires "the presence of a receptive and retentive soil."[25] The soil must be "receptive" in offering an environment that provides all the nutrients (starches and sugars) necessary for growth at the proper temperature. The soil must also be "retentive," in contrast to Luke 8:6, meaning it must be able to absorb rainfall or water from irrigation and retain that moisture long enough to get the seed "screwed into the ground."[26] The seed that "fell into [*eis*] the good soil" had the greatest chance of surviving and thriving because the ecological conditions for growth allowed the seed to flourish and produce an unexpected harvest.

Another important ecological perspective surfaces in this particular parable: the issue of threat to the seed ecosystem. Generally, one may classify an ecosystem threat as a *disturbance* (naturally occurring) or *perturbation* (human manipulations). Disturbances "focus primarily on distinct events that disrupt the function of ecosystems."[27] Natural disturbances include

[24]Murray J. Harris, *Prepositions and Theology in the Greek New Testament* (Grand Rapids: Zondervan, 2012), 29.

[25]Daniel Hillel, *The Natural History of the Bible: An Environmental Exploration of the Hebrew Scriptures* (New York: Columbia University Press, 2006), 146.

[26]F. Nigel Hepper, *Baker Encyclopedia of Bible Plants: Flowers and Trees, Fruits and Vegetables, Ecology* (Grand Rapids: Baker Books, 1992), 86.

[27]Paul Rogers, *Disturbance Ecology and Forest Management: A Review of the Literature* (Washington, DC: US Department of Agriculture, 1996), 1.

drought, fire, excessive wind, moving water, animal intrusion, and various types of disease. All of the threats or disturbances identified in this parable are natural and pose some danger to the growth of the seeds. All ecosystems are susceptible to either internal or external disturbances that threaten the well-being of individual organisms and the ecosystem as a whole. An ecosystem's ability to withstand disturbance is *resistance*, and the speed with which it is able to resist is *resilience*.[28] Both are enhanced by the diversity of an ecosystem because the greater the biodiversity, the more resources are available to fight a disturbance.

The seeds in this ecology did not encounter ideal environmental conditions. We see external threats from animals ("the birds of the air ate it up," Lk 8:5), from harsh soil conditions ("fell on rocky *soil*," Lk 8:6), from lack of water ("because it had no moisture," Lk 8:6), and from hostile elements of the landscape ("the thorns grew up with it and choked it," Lk 8:7). Yet despite these external threats to the seed, some "fell into the good soil," and the seeds "grew up, and produced a crop" (Lk 8:8). Here we have two verb synonyms for growth (grew and produced). They were perhaps deliberately selected to match the hostile soil environments and the description of the harvest. Most parable scholars direct their attention to the stark contrast between the harsh soil conditions and the abundant harvest. Some, like John Dominic Crossan, argue that the description of seed growth is foreign and anachronistic, insisting that Mark or some other scribal hand inserted this element into the original parable.[29] While the first group mines the contrast for valuable insights, they overlook the equal emphasis on growth of the seeds that settled in the hospitable soil. Without this intervening period of growth, there would be no harvest, regardless of size. Crossan, for some reason, finds it difficult to understand that between sowing and harvest transpires a period of growth. The verbs for growth emphasize the power of the soil ecology to produce flourishing seed. These are not seeds struggling for survival; they thrive because they have landed in an environment that is perfectly suited for growth.

---

[28]Robert Leo Smith, *Ecology and Field Biology*, 5th ed. (Menlo Park, IL: Addison Wesley Longman, 1996).

[29]John Dominic Crossan, "The Seed Parables of Jesus," *Journal of Biblical Literature* 92, no. 2 (1973): 244-66.

THE PARABLE OF THE MUSTARD SEED:
MATTHEW 13:31-32, MARK 4:30-32, AND LUKE 13:18-19

The parable of the mustard seed is one of the most commonly known of the parables Jesus told to explain the growth of the kingdom, and it is also one of the most controversial. The controversy stems from the mustard seed's description as the "smallest of all seeds" (Mt 13:32 NIV). The controversy is moot once we recognize, as does Hepper, that Jesus is no doubt referring to "the smallest of *cultivated* seeds" since Luke "seems to indicate a garden plant, not a wild one."[30] Mark Bailey confirms this interpretation by noting that "the verb *lachainō* means 'to dig,' and therefore relates to plants that are intentionally planted, rather than those that might grow wild."[31]

Since the parable of the mustard seed is another in a series of growth parables explaining the growth of the kingdom of God, we must pay attention to the ecological overtones in the parable. The point of the parable seems to be that the size of a seed does not preclude the potential for substantive growth. The mustard seed (*sinapi*) begins at about "one millimeter in diameter and weighing less than 1/700 of a gram."[32] However, it may reach a normal height of six feet with the potential to reach a height of fifteen feet.[33] The focus of the parable is on the vigorous growth from seedling to mature plant. This is apparently the point Jesus makes by contrasting the seed's initial miniscule size and its eventual height with spreading branches.

Mark mentions soil twice, whereas Matthew mentions a field, and Luke a garden (Mk 4:41; Mt 13:31; Lk 13:19). All three terms refer to soil ecology that describes the habitat where seeds interact with the natural environment. Soil ecology includes a diverse earthen community of bacteria, fungi, worms, mites, and other organisms. Nutrient pathways among the various components of a soil ecosystem provide for the mutual sharing of minerals, water, and carbohydrates. Earthworms contribute to soil ecology by burrowing to enhance water infiltration and aerate the soil. Fungi create nutrient links

---

[30]Hepper, *Bible Plants*, 133.
[31]Mark L. Bailey, "The Parable of the Mustard Seed," *Bibliotheca Sacra* 155 (1998): 453n20.
[32]James W. Scott, "The Misunderstood Mustard Seed," *Trinity Journal* 36, no. 1 (2015): 28.
[33]Kim-Huat Tan, "Mustard," *The New Interpreters Dictionary of the Bible* (Nashville: Abingdon, 2009), 4:183.

(fungal hyphae) between living organisms to share needed resources for growth.[34] Seeds planted in this rich soil ecology become a part of the underground ecosystem that contributes nutrients and absorbs what they need to germinate and produce a plant.

The emphasis on soil ecology in all three versions of the parable may suggest that we pay closer attention to this element of the parable. Mustard seeds, or any seeds for that matter, will not germinate, grow, and reproduce if they do not have the proper soil ecology, whether that soil is in a field or in a cultivated garden. Jesus wanted his disciples to grasp this critical component of natural and spiritual growth. The kingdom will grow from an insignificant beginning into an abundant harvest only if is situated within the proper spiritual ecology. Members of the kingdom will share in this growth only if they too exist within a spiritual ecology that provides the requisite spiritual nutrients. We learn from other New Testament texts that Jesus identifies himself, his Spirit, and his word as key elements of this spiritual ecosystem (Jn 14–15). We learn from the words of the apostle Paul that other Christians participate in this spiritual ecology of salvation (Eph 4:11-16). The parables of growth that Jesus taught prepared his disciples to receive further instruction that would illuminate their understanding of spiritual growth from an ecological perspective.

SUMMARY

Jesus taught in parables, and when he drew the starting point of his parables from features of natural growth in creation, he regularly identified features that one could classify as ecological. He identified features of natural growth that stressed the interconnected and interactive nature of growth in ecological settings. The seed parables we highlighted in this chapter partake of those ecological elements. The parable of the seed growing ecologically reminds us that ecological growth is sequential and invariant. The parable of the sower and three types of soil directs our attention to soil ecology and the environmental threats that have the potential to subvert successful growth. The parable of the mustard seed draws attention to its insignificant origin, which it is able to overcome with the help of a supportive ecosystem. All of

---

[34]Suzanne W. Simard et al., "Net Transfer of Carbon Between Ectomycorrhizal Tree Species in the Field," *Nature* 338, no. 7 (August 1997): 579-82.

the parables share a bountiful end state that arrests the attention of observers: "the harvest has come," "produced a crop a hundred times as great," and "it is larger than the garden plants." In the next chapter, we will use some of the observations from these parables to inform our understanding of ecological growth in the body of Christ.

— FOUR —

THE ECOLOGY OF THE BODY OF CHRIST

To be a member is to have neither life, being, nor movement,
except through the spirit of the body, and for the body.

BLAISE PASCAL, *PENSEÉS*

A few years ago, *Forbes* magazine published an article titled "How to Train Like an Olympian."[1] In the article they describe the regimented and demanding process that world-class athletes must follow if they are to achieve Olympic glory. Most of our Olympic champions invest four to eight years training for a few minutes of competition. All of them must submit to annual training regimens supervised by coaches, nutritionists, trainers, dieticians, physicians, and exercise physiologists. Surprisingly, they must be willing to sleep eight to ten hours every night and nap between thirty and ninety minutes every afternoon so that their bodies rebuild tissues and muscles broken down during their rigorous training program. The results are athletes like Carl Lewis, Lindsey Vonn, Michael Phelps, and Amanda Beard. The human body responds well to exercise and training because it is being asked to utilize all of the complementary bodily systems in a synergistic symphony aimed at peak performance. The Creator perfectly designed the ecology of the human body to strengthen and grow in response to proper exercise and nutrition, and thus body ecology serves as an apt illustration for the church as the body of Christ.

---

[1]"How to Train Like an Olympian," *Forbes*, July 8, 2008, www.forbes.com/2008/07/08/training -perfect-athlete-olympics08-forbeslife-cx_avd_0708health.html.

Howard Snyder, more than any other contemporary Christian theologian, recognizes the power of an ecological perspective of the church and kingdom.[2] Although he makes many significant contributions to our understanding of ecology from a biblical perspective, his greatest contribution comes from his application of ecological insights to our understanding of the church as the body of Christ. He writes, "The church is a socio-spiritual organism. Its life may be viewed ecologically as a dynamic interplay of several parts."[3] He then identifies biblical patterns that support this view of the church: "Paul presents one ecological view of the church in 1 Corinthians 12, using the analogy of the human body. . . . Paul speaks similarly but from a slightly different angle in Ephesians 4:1-16. This is also a picture of church ecology."[4] Concluding his argument he writes, "Examining Ephesians and other New Testament passages, we may construct an ecological model which traces the basic elements of the church's life and shows how these relate to each other."[5] While Snyder provides a helpful starting point for a biblical understanding of the church, we want to expand his original insight and apply it more precisely to the spiritual growth of God's people as members of the body of Christ. We are particularly interested in not only honoring the ecological reality of church life in general, but recognizing that this same ecological dynamic governs community formation and individual Christian growth.

THE BODY OF CHRIST

There are some who believe that evangelical scholarship has not given as much attention to the concept of the body of Christ as an organizing concept as the analogy requires. F. LeRon Shults surveys the concept among evangelical biblical and systematic theologians and concludes, "When biblical images of the church are examined by evangelical scholars, the body of Christ is almost always listed but does not serve a vibrant or integrative

---

[2]Howard A. Snyder, *Liberating the Church: The Ecology of Church and Kingdom* (Downers Grove, IL: InterVarsity Press, 1983); *Radical Renewal: The Problem of Wineskins Today* (Eugene, OR: Wipf and Stock, 1996); *Decoding the Church: Mapping the DNA of Christ's Body* (Grand Rapids: Baker Books, 2002); *Salvation Means Creation Healed: The Ecology of Sin and Grace* (Eugene, OR: Cascade Books, 2011).

[3]Snyder, *Liberating the Church*, 74.

[4]Ibid.

[5]Ibid., 74-75.

function."[6] When N. T. Wright identifies those who have made the body of Christ "central" in their theology, he only cites E. P. Sanders and Albert Schweitzer.[7] Of course, there are exceptions to this general description.[8] James D. G. Dunn argues that the Pauline image of the church as body of Christ "is the dominant theological image in Pauline ecclesiology."[9]

Paul used the human body to illustrate the dynamic interconnection between Christ and his church, as well as the organic interconnections between and among Christians (1 Cor 12:27; Eph 4:16). As Ridderbos says, "For believers do not together constitute one body because they are members of one another, but because they are members of Christ, and thus are one body in him."[10] The connection of the body to its head is an obvious emphasis in Paul's discussion of this theme. The organic connection Christians have with Christ as head of the body establishes an important Pauline theme of our distinctive relationship "in Christ" (*en Christō*) and "with Christ" (*syn Christō*) that defines our new identity as "a new creature" (2 Cor 5:17).

The concept of the church as the body of Christ (*sōma Christou*) rests upon a base set out by Gosnell Yorke. He writes, "*Sōma*, with regard to the church, has the human body as its metaphorical referent," and further, "The human *sōma* and not Christ's *sōma* is used consistently as the term of comparison for the church as *sōma*."[11] Paul does with the body concept what Jesus does with the parables of growth; both use an example from nature to illustrate a spiritual reality. Jesus illustrated kingdom growth by reference to identical processes of growth in flowers and seeds. Paul illustrated the functioning and growth of the body of Christ by reference to the functioning and growth of the human body. It becomes important, therefore, to understand the literal referent Paul employs to make his spiritual application to the relationship between Christ and his church. One of the glaring omissions in the biblical and theological treatments of the body of Christ in Paul is to

---

[6]F. LeRon Shults, "The 'Body of Christ' in Evangelical Theology," *Word and World* 22, no. 2 (2002): 178-85, 181.

[7]N. T. Wright, *Paul and the Faithfulness of God* (Minneapolis: Fortress, 2013), 825.

[8]Ernest Best, *One Body in Christ* (London: SPCK, 1955); Gosnell L. O. R. Yorke, *The Church as the Body of Christ in the Pauline Corpus* (Lanham: University of America Press, 1991); John A. T. Robinson, *The Body: A Study in Pauline Theology* (London: SCM Press, 1952).

[9]James D. G. Dunn, *The Theology of Paul the Apostle* (Grand Rapids: Eerdmans, 1998), 548.

[10]Herman Ridderbos, *Paul: An Outline of His Theology* (Grand Rapids: Eerdmans, 1997), 376.

[11]Yorke, *Pauline Corpus*, xv, 10.

move almost immediately to spiritual application without first exploring the original human image from which Paul draws his spiritual implications.[12]

THE HUMAN BODY AS ECOSYSTEM

Paul understood through divine revelation what we now understand through medical science about the ecological functioning of the human body. It is commonplace in contemporary medical circles to refer to the human body as an ecosystem because all of the various parts and individual functions interact with each other in a mutualistic symphony for the benefit of the body as a whole. Daniel Schneck flatly states, "The human body is an ecosystem."[13] Schneck then elaborates upon his own terse comment: "At the microscopic scale of perception, then, by definition, the interior of the human body fits the description of an ecosystem."[14] The June 2012 cover of *Scientific American* had an impressionistic drawing of the human body with the headline, "Your Inner Ecosystem." The opening sentence in the lead article states the concept succinctly: "Over the past 10 years or so, however, researchers have demonstrated that the human body is not such a neatly self-sufficient island after all. It is more like a complex ecosystem."[15]

The human body, as a well-functioning ecosystem, engages in a constant exchange of nutrients and other necessary resources through the circulatory system. All of the various subsystems of the human body (immune, respiratory, nervous, skeletal, muscular, and endocrine) constantly engage in homeostatic biofeedback interaction and communication in order to maintain a quasi-equilibrium that sustains health and life. That is, the body is constantly monitoring itself across all of its ecological systems to maintain proper balance and stability. All of the system functions of the human body lead scientists to recognize the ecological character of these interactions. Among other things, Paul highlights this feature of how the human body functions and sets it parallel to how the body of Christ functions. As Ernst Käsemann writes, "The apostle uses the expression 'the

---

[12]While scholars address the origin of Paul's body concept, we are more concerned about what Paul saw in how the human body functions than in the origin of the term.

[13]Daniel J. Schneck, "What Is This Thing Called 'Me'? The Stationary, Buffered, Isothermal Living Engine," *American Laboratory* 38, no. 10 (2006): 4.

[14]Ibid.

[15]Jennifer Ackerman, "The Ultimate Social Network," *Scientific American*, June 2012, 38.

body of Christ' because he really means to point out the structural characteristics of a body."[16] Although many Pauline scholars speculate about the background of the body (*sōma*) imagery, Grenz makes Käsemann's point when he notes that "the background for this picture does not lie so much in the Old Testament as in human anatomy."[17]

We can also point to the cultural influence of the city of Corinth as another possible reason why Paul employs the body imagery in 1 and 2 Corinthians.[18] In the city of Corinth stood a temple dedicated to the worship of Asclepius, the god of disease and medicine. The pagan worshiper would bathe in the sea and enter the temple precincts, where they would undergo ritual purification. Upon completion of these initial rites, worshipers would then be ushered into the *abaton*, where they enjoyed a warm bath and were provided a bed to spend the night. During sleep, Asclepius would either heal the person or reveal a cure that would offer healing if performed by the worshiper. When the individual experienced healing, they would take a votive terracotta replica of the healed body part to the temple, and the temple attendants placed it on display in a room set aside for expressing gratitude to the god.

According to the lead excavator of the temple site, there were some ten cubic meters of votive terracotta body parts, representing almost every part of the body.[19] The excavations revealed 125 terracotta hands, in addition to numerous legs, knees, feet, fingers, ears, eyes, female breasts and reproductive organs, and male genitalia. The discovery of these life-sized body parts exceeds any discovered among the ruins of other Asklepieia in ancient Greece. Apparently there existed a thriving business in Corinth, known for terracotta work, that reproduced the body parts for grateful worshipers. Carl Roebuck says the cult of Asclepius "seems to have played its part in the life of Corinth for eight centuries, from its foundation in

---

[16]Ernst Käsemann, *Perspectives on Paul* (Philadelphia: Fortress, 1971), 104.

[17]Stanley J. Grenz, *Theology for the Community of God* (Grand Rapids: Eerdmans, 1994), 466.

[18]Andrew E. Hill, "The Temple of Asclepius: An Alternative Source for Paul's Body Theology?," *Journal of Biblical Literature* 99, no. 3 (1980): 437-39.

[19]Professor F. J. De Waele, who served on the staff of the American School of Classical Studies at Athens, guided the original excavation of the temple site from 1929 to 1934 when his work was interrupted by the events leading to the outbreak of World War II. Carl Roebuck, in 1951, published De Waele's original work and added his own subsequent work on the site to which De Waele expressed his general agreement.

the late fifth century BC until its destruction in the late fourth century after Christ."[20]

The emphasis and almost exclusive concern of the Asclepian cult focused on the healing of individual body parts. 1 Corinthians is the only place in Paul's epistles where he mentions the body of Christ that identifies specific body parts like the foot, hand, ear, eye, and head (1 Cor 12:15-21).[21] However, Paul wants the Christians in Corinth to understand that the human body is not just a collection of individual body parts but is also an integrated totality that functions as a whole body. He wants Christians to see the interconnected and interdependent relationship of individual body parts to the whole body. The Corinthians understood the diversity of the human body; they needed also to understand its unity ("But now there are many members, but one body," 1 Cor 12:20). This is the emphasis in the often-overlooked prefatory comments Paul makes in 1 Corinthians 10:16-17 ("Since there is one bread, we who are many are one body; for we all partake of the one bread"). The mutuality of parts to whole in a defined ecosystem such as the human body expresses Paul's understanding of how the body of Christ functions to ensure a vital, thriving, and growing organism.

The ecosystem-like functioning of the human body is at the heart of what Paul conveys in Romans 12, 1 Corinthians 12, and Ephesians 4. The "one" and "many" language Paul uses in 1 Corinthians 12, as well as the whole body/every joint contrast in Ephesians 4:16, support an ecological view of the body of Christ. Dunn, although not referring to the body imagery in Paul as specifically ecological, recognizes that Paul's "point of the image" expresses "mutual interdependence of its members."[22] Schnelle can see that members of the body "are not all the same, but all are interconnected and need each other."[23] Gordon Fee also observes that Livy, who used the body metaphor, stressed "the *interdependence* of the many on one another" and that Seneca was concerned to

---

[20]Carl Roebuck, *Corinth: The Asklepieion and Lerna* (Princeton, NJ: The American School of Classical Studies at Athens, 1951), 159.

[21]Paul refers to body of Christ in Rom 7:4; 12:4-5; Eph 1:22-23; 4:12, 16; 5:23; Col 1:18, 24; 2:19.

[22]Dunn, *Theology*, 555. Dunn makes multiple references to this primary characteristic of Paul's concept of the body of Christ. In regard to what he calls "body imagery" the point of the imagery is the "interactive relations of the worshiping community in general" (550), "in mutual relations . . . mutual cooperation . . . mutual interdependence . . . and mutual responsibility" (552), "the mutual interdependence of its members" (555).

[23]Udo Schnelle, *Apostle Paul: His Life and Theology* (Grand Rapids: Baker Academic, 2003), 564.

show "that the parts are mutually interdependent."[24] The ecological language of mutual interdependence and interconnection employed by these writers to describe the function of the human body would fit into any scientific description of natural ecology. For instance, Fritjof Capra identifies three basic principles of ecology, the first of which is interdependence—defined as "the mutual dependence of all life processes on one another," which, he insists, "is the nature of all ecological relationships."[25] Donald Worster cites one ecologist who observed, "Everything is hooked to everything else—not superficially, as in a machine, but essentially, as in the human body."[26]

THE BODY OF CHRIST AS ECOSYSTEM

Ever since the church renewal movement in the 1960s, American Christians are accustomed to recognizing the New Testament picture of the church as a living, dynamic organism. However, as Howard Snyder says, there is "a key question for every organism: What is its ecology?"[27] This question reveals what has been a missing component in a fully orbed understanding of the church as an organic community. While we have rightly recognized the organic dimension in Paul's concept of the church as the body of Christ, we have forgotten that all living organisms can only live, grow, and reproduce within a defined ecosystem. Living organisms need a diverse ecosystem within which they connect and interact. Jesus highlighted this basic law of natural growth through his parables. Everything God created grows ecologically, including the kingdom, the church, and the Christian, since, as Hunter says, the patterns of growth that are "valid in one sphere" are "valid also in the other."[28] Our interest in the church as the body of Christ is restricted to understanding how the church as the body of Christ grows both from an individual and corporate perspective.

In chapter two, we noted the etymology of the word *ecology* and its use by the pioneers in the field to refer to "the science of the household of

---

[24]Gordon D. Fee, *The First Epistle to the Corinthians* (Grand Rapids: Eerdmans, 1987), 602n12. Emphasis in text.

[25]Fritjof Capra, *The Web of Life* (New York: Anchor Books, 1997), 298. The other two basic principles are flexibility and diversity (302-4).

[26]Donald Worster, *Nature's Economy: A History of Ecological Ideas*, 2nd ed. (New York: Cambridge University Press, 1994), 317.

[27]Snyder, *Radical Renewal*, 117.

[28]Hunter, *Interpreting the Parables*, 8.

nature."[29] The words *oikia/oikos* and their derivatives in Greek all have something to do with the household of humans who inhabit a particular physical and social space (Mt 8:14; Acts 10:2, 22, 30). We often refer to the original location of the gathered community as "house churches" because the church met in a literal house (see Rom 16:5; 1 Cor 16:19; Col 4:15). New Testament writers like Paul utilized the semantic field of *oikos* to express their understanding of the nature and mission of the church.[30] For instance, Paul and Peter referred to leaders in the church as "stewards" or "managers" (*oikonomoi*; see 1 Cor 4:1-2; Gal 4:2; 1 Pet 4:10). Paul referred to God's missional plan for the church as an *oikonomia* (Eph 1:10; 3:2).[31] When Paul described corporate or individual spiritual growth as "building up" or "edification," he used *oikodomē* or *oikodomeō* (see Rom 14:19; 15:2; 1 Cor 14:3, 5, 12, 26; Eph 4:12, 16, 29; 1 Thess 5:11). When Paul referred to the church as "God's building" he again used *oikodomē* (see 1 Cor 3:9; Eph 2:21). Finally, when Paul wanted to refer to the church as a family, he described it as "the household [*oikeios*] of the faith" (Gal 6:10) or "God's household [*oikeios*]" (Eph 2:19).

In Ephesians 2:19-22, Paul used six different *oikos*-related terms as he contrasted the previous status of Gentiles to their new status in Christ. We have put the *oikos*-derived terms in parentheses after the English translation and you can identify the root by noting the appearance of *oik* somewhere in the Greek word:

> So then you are no longer strangers and aliens [*paroikoi*], but you are fellow citizens with the saints, and are of God's household [*oikeioi*], having been built on [*epoikodomēthentes*] the foundation of the apostles and prophets, Christ Jesus Himself being the corner *stone*, in whom the whole building [*oikodomē*], being fitted together, is growing into a holy temple in the Lord, in whom you also are being built together [*synoikodomeisthe*] into a dwelling [*katoikētērion*] of God in the Spirit.

All of this *oikos* language in Ephesians 2 hearkens back to Ephesians 1:10, where Paul describes the carrying out of the mission of the church as the

---

[29]Astrid Schwarz and Kurt Jax, eds., *Ecology Revisited: Reflecting on Concepts, Advancing Science* (Dordrecht: Springer Science, 2011), 145.

[30]Roger W. Gehring, *House Church and Mission: The Importance of Household Structures in Early Christianity* (Peabody, MA: Hendrickson, 2004).

[31]See Timothy A. van Aarde, "The Use of *Oikonomia* for Missions in Ephesians," *Verbum et Ecclesia* 37, no. 1 (2016): a1489, http://dx.doi.org/10.4102/ve.v37i1.1489.

oikonomia (missional plan) of God. Part of that plan is to bring together believing Jews and Gentiles into "one new humanity out of the two, thus making peace" between the previously antagonistic communities (Eph 2:15 NIV).

One cannot fully interpret or understand Paul's message in this passage about the relationship between Christian Jews and Gentiles in the body of Christ without also taking into consideration his repeated use of *syn*-compound language in this same text. Using the same textual layout we used above to highlight the *oikos*-related terms, we will now highlight the three *syn*-compounds formed with the prefix *syn/sun/sum* ("together with") plus a verb or noun in Ephesians 2:19-22:

> So then you are no longer strangers and aliens, but you are fellow citizens [*sumpolitai*] with the saints, and are of God's household, having been built on the foundation of the apostles and prophets, Christ Jesus Himself being the corner *stone*, in whom the whole building, being fitted together [*synarmologoumenē*], is growing into a holy temple in the Lord, in whom you also are being built together [*synoikodomeisthe*] into a dwelling of God in the Spirit.

The powerful rhetorical combination of the *oikos*-rooted words and the *syn*-compounds give the impression that Paul understands the church as something built by the Godhead (Father, Son, and Spirit, Eph 2:18, 20, 22) working in divine collaboration to perfectly construct "a dwelling of God in the Spirit." This is a structure (*oikos*) in which God fits and builds together (*syn*) all of the disparate living materials (believing Jews and Gentiles) to form a cohesive and highly interconnected spiritual entity that will be able to carry forward the *oikonomia*—God's plan of salvation and sanctification for the world (Eph 1:10; 3:2). This *oikonomia*, which we understand as God's mission for the world, Paul links to the "mystery of Christ," which was disclosed to Paul, and reveals, namely, "that the Gentiles are fellow heirs [*synklēronomos*] and fellow members of the body [*syssōmos*], and fellow partakers [*symmetochos*] of the promise in Christ Jesus through the gospel" (Eph 3:4, 6). Here Paul clusters his *syn*-compounds to reinforce the equal status of believing Gentiles with believing Jews in the church. The repeated English word *fellow* appears in translations in order to convey the force of the *syn* prefix in each of the adjectives Paul employed in this verse. As Harold Hoehner says, the mystery to which Paul alludes is not that Gentiles would be saved "but rather that believing Jews and Gentiles are together in Christ."[32] And this is not only togetherness as unequal parts but togetherness as equals, since that is obviously the force of the *syn* prefix in each of these three *syn*-compounds. As Robinson writes, "Time and again, he coins strange new words with the prefix *syn* rather than use the plain preposition. He clearly feels the inadequacy of language to convey the unique 'withness' that Christians have in Christ."[33] To compensate for this "inadequacy of language," Paul combines the prefix *syn* with a host of different nouns and verbs to convey the power of the gospel and the power of Christ's work on our behalf.

We highlight this emphasis in Paul because he builds upon it to teach that the body of Christ grows ecologically as all of the interconnected parts of the body function harmoniously and interactively. This is what Ernest Best gets at when he writes:

> If one member fails to exercise his gift, that hinders both the growth of the whole and the individual growth of each member (no member can grow apart from the whole). The growth of each member is involved in the growth of the whole and the growth of the whole in the behavior of each member.[34]

Best can take this interactive view of the nature of growth in the body of Christ because he recognizes that "there is some kind of identity between the believer and Christ . . . he is connected to Christ by a bond similar to that which links together two parts of an organism."[35] The connection between believer and Christ creates a means whereby the riches of Christ transfer to each connected Christian. Since each believer (Jew and Gentile) has a connection to Christ, each has a connection to other members of his body. In order for this to happen, the connections need to be intact and provide a conduit for the exchange of spiritual resources. The ideal description of this properly functioning ecosystem appears in Paul's vivid portrait in Ephesians 4:16:

> From whom the whole body, being fitted [together] [*synarmologoumenon*] and held together [*symbibazomenon*] by what every joint supplies, according

---

[32]Harold W. Hoehner, *Ephesians: An Exegetical Commentary* (Grand Rapids: Baker Academic, 2002), 448.

[33]Robinson, *Body*, 63.

[34]Best, *One Body*, 187.

[35]Ibid., 55.

to the proper working of each individual part, causes the growth of the body for the building up of itself in love.

We see this same *syn* language in Colossians 2:19: "From whom the entire body, being supplied and held together [*symbibazomenon*] by the joints and ligaments [*syndesmōn*], grows with a growth which is from God."

Both passages depict connections ("fitted and held together") that the different parts of the body ("joints and ligaments") have to each other to form a network that holds the body together. Both passages add the element of reciprocal exchange between the various connections ("by what every joint supplies," "being supplied," and "by the joints and ligaments"), which teaches us that having network connections is necessary but not sufficient to produce growth. If the connections did not "supply" nutrients and nourishment to all of the interconnected parts of the body, there would be no growth. The static connections need dynamic exchanges of nutrients to produce "the growth of the body." This is the ecology of the human body at work and Paul sees in this an illustration of how the ecology of the spiritual body of Christ functions to produce spiritual growth.

Thinking of the church as ecosystem reflects how Paul and others applied ecological concepts, such as mutual interdependence, interconnectedness, organic interaction, and diversity, to the body of Christ.[36] While scholars have observed and identified these characteristics of the body of Christ from Scripture, they have given little attention to the spiritual growth implications for how the body of Christ grows corporately or how individual Christians, as members of the body, grow within the spiritual ecology of the body of Christ.

The organic connection between head and body makes possible and creates highly dynamic connections between and among individual members of the body that serve as interactive conduits for the exchange of spiritual nutrients. This means Paul understood the connections between members not simply as a static reality to contemplate but as a dynamic reality to activate. The static connections between members created by our common connection to Christ are also conduits between members, which allow them space to share spiritual resources they received from Christ in a

---

[36]Snyder, *Creation Healed*, 214-15; Dunn, *Theology*, 551; Christian Schwarz, *Paradigm Shift in the Church* (Carol Stream, IL: ChurchSmart Resources, 1999).

mutualistic manner (Rom 12:5; 1 Cor 12:25; Eph 4:16; Col 2:19). Christians activate these spiritual connections between members of the body by mutual exchanges of spiritual nourishment—seen primarily, but not exclusively, in the "one another" imperatives—empowered by the spiritual gifts distributed freely from the Holy Spirit (Rom 12:6; 1 Cor 12:7, 11; 14:12).

We know from the study of the ecology of human development and from other interactional developmentalists that humans develop and grow within a social ecology through reciprocal interactions, transactions, and exchanges.[37] The highly dynamic human environment instigates growth in the same way that Solomon observed: "As iron sharpens iron, so one person sharpens another" (Prov 27:17 NIV). This is what Karl Barth suggests as his understanding of how the church as the body of Christ experiences edification. He writes, "It is not the Christian individual as such, but the community which, in its individual members and through their reciprocal ministry, is edified, and lets itself be edified, and edifies itself."[38] Christians in the body of Christ sharpen one another spiritually through "reciprocal ministry" and through various engagements, interactions, relationships, experiences, and encounters that comprise the rich ecology Christ provides through the ministry of the Holy Spirit. We should note that the references to growth in these passages follow upon the interactive description of the ecology of the body of Christ. Individual growth can only occur within a defined ecology, and that is as true of the ecology of the church as it is for other parts of creation. When Christians disengage from or disrupt the ecological exchanges ("reciprocal ministry") taking place in the body of Christ, they short-circuit God's design for growth.

For this reason, Paul stresses the unity of believing Jews and Gentiles in Ephesians, is so concerned about the factions and schisms at work in the church in 1 Corinthians, and admonishes Euodia and Syntyche to overcome their differences in Philippians. A divided body of Christ disconnects these

---

[37]See Urie Bronfenbrenner, *The Ecology of Human Development* (Cambridge, MA: Harvard University Press, 1979); R. M. Lerner, *Concepts and Theories of Human Development* (Mahwah, NJ: Lawrence Erlbaum Associates, 2002); D. Magnusson and V. L. Allen, *Human Development: An Interactional Perspective* (New York: McGraw Hill, 1983); Lev Vygotsky, *Thought and Language* (Cambridge, MA: MIT Press, 1986); Arnold Sameroff, ed., *The Transactional Model of Development* (Washington, DC: American Psychological Association, 2009).

[38]Karl Barth, *Church Dogmatics* XIV, §15.67, 627.

vital corridors of nutrient exchange and interferes with the spiritual growth process. We think this is why Paul must also address the concomitant issue of spiritual immaturity in the church at Corinth. Paul addresses this directly in 1 Corinthians 3:1-4 when he links their spiritual immaturity ("infants in Christ") with their divisions ("for . . . there is jealousy and strife among you"). Anytime we experience a disruption in our relationships with other Christians, we jeopardize the spiritual growth of all the parties involved. When this situation encompasses the entire church, we require intervention on the part of church leaders to seek restoration and reconciliation. Without some kind of intervention, the spiritual vitality and witness of the congregation remains in jeopardy.

SUMMARY

The dynamic ecology of the body of Christ creates the perfect environment for spiritual growth. When the spiritual connections and reciprocal interactions are disturbed through external or internal threats, intentional intervention is needed to restore the spiritual equilibrium of God's spiritual ecology. An ecological perspective of the body of Christ helps us to see God's design and accommodate our educational endeavors and church ministries accordingly. Through the lens of the ecology of the human body, we are better able to understand how God intends the spiritual ecology of the body of Christ to grow so that we can maximize our own individual growth and the collective growth of the Christian community. When functioning properly through spiritual connections and nutrient exchanges, the body of Christ "grows with a growth which is from God" (Col 2:19). In the next chapter, we will apply these ecological principles to online learning environments where Christians gather to study and learn together.

So far in part one, we have seen that the universe God created operates at all levels in an ecological fashion built upon the twin features of interconnection and interaction. The interfaced connections and interactions within an ecosystem may be organic, atomistic, social, or spiritual. Living organisms in natural ecologies are interconnected and interactive and likewise human beings in social ecologies are interconnected and interactive. The nature parables that Jesus taught and the metaphor of the body of Christ that Paul used partake of these natural and social ecological components. For kingdom growth to occur or for the body of Christ to grow to full maturity, functioning spiritual connections and interactions must sustain growth individually and collectively. In part two, we move from this conceptual examination of ecology to praxis within the defined learning ecology of Christian higher education—with special attention to online education. Here we will tease out specific ways in which we can design, structure, and implement the learning ecology of an online environment to mirror the ecological components of interconnections and interactions.

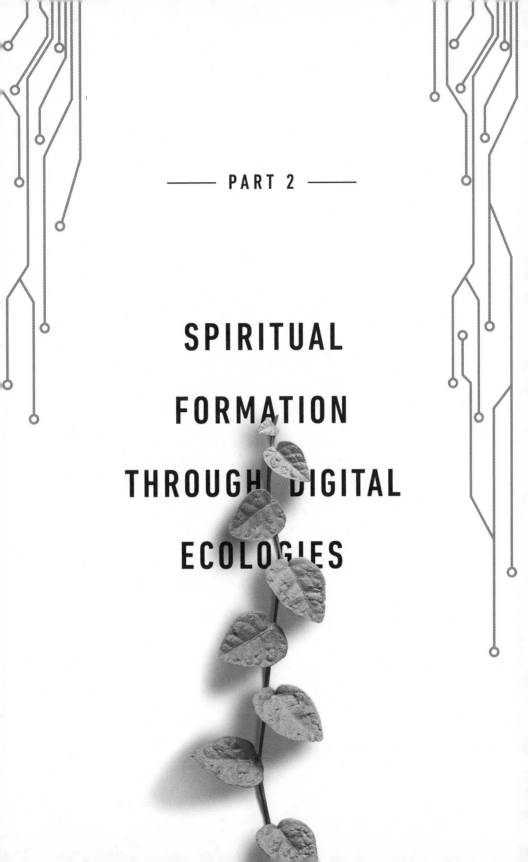

PART 2

SPIRITUAL

FORMATION

THROUGH DIGITAL

ECOLOGIES

— FIVE —

GROWING TOGETHER ONLINE

People started telling me that they felt connected in some kind
of community sense. I used to be doctrinaire about definitions
and I didn't feel it was a community site, but I eventually
said, if people feel connected, it must be a community.

ROBERT PUTNAM, *BETTER TOGETHER*

As we move into the heart of our case for an ecological perspective on spiritual formation, particularly as it applies to online learning, it is important to mark out our road map through the learning ecologies we will navigate. In this next section, we will apply our ecological model to the creation of online learning ecologies, taking into consideration insights from the field of social network science and the larger digital landscape we currently inhabit.

When we hear the term *capital*, we often think in terms of financial assets or the financial worth and value of those assets. We can also apply this term to social networks and the value of reciprocal relationships within those networks. There is a certain inherent value for an individual or group of individuals who receive this social capital. For example, a dear friend of ours was deployed to Iraq as a military chaplain for eighteen months. During his absence, his family and friends were able to communicate not only in letters and packages but also through Skype in real time. These virtual encounters, mediated by the available technology, produced a social capital or benefit that was different from simple one-way forms of communication. Friends and family were able to see him, hear him, observe his living quarters, and get a real sense of what life was like for him so far from home. Although not physically present with each other, these social interactions provided valuable benefits to the chaplain, his family, and his friends.

Robert Putnam and Lewis Feldstein maintain that "social capital refers to social networks, norms of reciprocity, mutual assistance, and trustworthiness. . . . Social networks have real value both for the people in those networks . . . as well as for bystanders."[1] When we look at social networks from the perspective of ecosystems, we can better understand their worth or value in that both the recipient and the donor create mutual benefit or assistance. It isn't that there is simply mutual assistance or benefit in social networks; connectivity between the various components in the network forms the basis for how the network expresses its identity and how it grows and develops.

ECOLOGICAL COMMUNITIES

Most ecologists understand that normal, healthy growth results from being in the right environment. We know from observations of nature that organisms thrive because of a confluence of factors that contribute to the overall development of a particular species. Commenting on the relationship between individual and ecological growth, Alfred North Whitehead—scientist, philosopher, and mathematician of the late 1800s and early 1900s—wrote:

> You may obtain individual specimens of fine trees either in exceptional circumstances, or where human cultivation has intervened. But in nature the normal way in which trees flourish is by their association in a forest. Each tree may lose something of its individual perfection of growth, but they mutually assist each other in preserving the conditions for survival. A forest is the triumph of the organization of mutually dependent species.[2]

One of the most fascinating examples of this is California's redwood forests. The trees are massive giants, and yet their root system is relatively shallow. The critical component to the trees' survivability is their mutual

---

[1] Robert Putnam and Lewis Feldstein, with Don Cohen, *Better Together: Restoring the American Community* (New York: Simon & Schuster, 2003), 2.

[2] Alfred Whitehead, *Science and the Modern World* (Cambridge: Cambridge University Press, 1953), 257. While we find Whitehead's scientific observations pertinent to our discussion about ecology, we do not endorse his ventures into theological speculation regarding process theology.

dependence on one another. Other trees stabilize them during periods of turbulence. Brian Henning and Adam Scarfe add, "Every entity, or phenomenon studied, requires its environment in order to exist, and every environment is comprised by an indefinite multitude of interdependent factors, each conspiring to bring about what is."[3]

Social scientists might argue similarly—that humans grow and develop best when we are socially connected with other growing humans. In fact, Andrew Feenberg and Maria Bakardjieva point out that human community fosters or instigates "a large share of human development."[4] Social scientists such as Urie Bronfenbrenner, Richard Lerner, and David Magnusson maintain that normal human development is sustained and encouraged in the reciprocal exchanges we have with other people.[5] It isn't simply, as Bronfenbrenner described, that we exist in certain concentric circles of relationship to one another; rather, in order to grow and develop in a way that meets certain norms or standards, there must be a degree of reciprocity and interconnection between and among our circles of connection. Bronfenbrenner argued that the "phenomenon of interaction is of course fundamental to an understanding of how human beings develop."[6] Just as individuals grow best when there is interaction in relationships, so do communities experience healthy growth and development when there are reciprocally beneficial exchanges that "mutually assist each other."[7] Community interaction, then, is larger than a collection of individuals in that it forms and thrives with the relationships the individuals have to one another.

Most of us understand community to be comprised of individuals or groups who share particular interests or goals. Amitai Etzioni offers a two-pronged perspective of community by noting that it is "a web of affect-laden relationships among a group of individuals, relationships that often

---

[3]Brian Henning and Adam Scarfe, *Beyond Mechanism: Putting Life Back into Biology* (Lanham, MD: Lexington Books, 2013), 45.

[4]Andrew Feenberg and Maria Bakardjieva, "Consumers or Citizens? The Online Community Debate," in *Community in the Digital Age: Philosophy and Practice*, ed. Andrew Feenberg and Darin Barney (Lanham, MD: Rowman & Littlefield, 2004), 1-30.

[5]Urie Bronfenbrenner, *The Ecology of Human Development* (Cambridge, MA: Harvard University, 1979); Richard Lerner, *Concepts and Theories of Human Development* (Mahwah, NJ: Lawrence Erlbaum Associates, 2002); David Magnusson and Vernon Allen, *Human Development: An Interactional Perspective* (New York: Harcourt Brace Jovanovich, 1983).

[6]Bronfenbrenner, *Ecology of Human Development*, 146.

[7]Whitehead, *Science*, 257.

crisscross and reinforce one another . . . and a measure of commitment to a set of shared values, norms, and meanings."[8] Some communities form for a particular purpose or task, while others form over a lifetime. Some observers of online and in-person interaction question the extent or even legitimacy of communities formed by people who never engage one another face-to-face. Others, however, view the purpose and task of forming community as being possible both online and in person, acknowledging the significant impact both can have on group members. Feenberg and Bakardjieva would likely affirm the second position: "By online community, we mean relatively stable, long-term online group associations mediated by the Internet or a similar network."[9] Online students, for example, come together for a specific period of time to accomplish a particular purpose; in that process, they form a certain community of persons around various forms of presence. Whether the formation of community works itself out online or in person, it could be argued that community does not need to be forced as much as facilitated.

An ecological perspective of community integrates a number of interconnected variables. As in a healthy natural ecology, there are elements in social ecologies that work together to produce form and function. Julie Lytle, in describing ecologies of faith that form people, identifies those relationships or components as "interpersonal, social, cultural, economic, political and religious."[10] She adds that those various components can contribute to either a rise or decline in a certain equilibrium by what she identifies as "states of growth, stability, or decline."[11] There is a certain degree of homeostasis that ecosystems strive to maintain, and in this sense, communal ecologies should work to achieve a similar state, even though we know it will not last. An ecosystem view of community recognizes the multiple layers of influence on that system. Regardless of the venue, students who enroll in online courses or degree programs bring their personal ecosystem to bear

---

[8]Amitai Etzioni, "On Virtual, Democratic Communities," in *Community in the Digital Age: Philosophy and Practice*, ed. Andrew Feenberg and Darin Barney (Lanham, MD: Rowman & Littlefield, 2004), 225-38.

[9]Feenberg and Bakardjieva, "Consumers or Citizens," 2.

[10]Julie Lytle, *Faith Formation 4.0: Introducing an Ecology of Faith in a Digital Age* (New York: Morehouse, 2013), 118.

[11]Ibid.

on the learning experience. They represent communities like a family, a church, a neighborhood, or a group of friends. There is a bidirectionality between the influence of members in one's community and how those factors shape formation. There is a constant fluidity at play between components of one's personal ecosystem in which various parts serve to comprise the good of the whole. Growing together online requires the interdependence and interaction between the self and the communities in which one lives, serves, and learns.

GROWING ONLINE

While there are those who argue that digital venues undermine traditional forms of community and the leader's role of promoting spiritual development, others view the activities of online participants as an indicator of the desire and intention to further their growth. Heather Glover asserted, "Among the growing numbers of Americans who are using the Internet, many of them are turning to the digital dimension to get in touch with God and pursue matters of faith."[12] This mirrors what Elena Larsen and others at the Pew Research Center found in their work: "Fully 20% of Americans said they had shared their religious faith on social networking websites or apps (such as Facebook and Twitter) in the past week, and 46% said they had seen someone else share 'something about their religious faith' online."[13]

Some differentiate between online and real life as if to suggest that the former is distinctively different from the latter in form and function. The ecosystem paradigm we employ suggests that the networks we inhabit, whether online or off, are part of a larger ecological reality. There are critics of virtual, mediated, and online communities both inside and outside the Christian community. Since we are writing to those within the Christian community, we will limit our remarks to Christian critics of online community.

In the late 1990s, many evangelical seminaries accredited by the Association of Theological Schools (ATS) were experimenting with various forms of distance or virtual education using the latest technological innovations.

---

[12]Heather Glover, "Using the Internet to Renew Community Ties in Traditional Religious Communities" (PhD diss., Baylor University, 2003). Masters Abstracts International.

[13]Elena Larsen, "Cyberfaith: How Americans Pursue Religion Online," Pew Internet and American Life Project, December 23, 2001, www.pewinternet.org/2001/12/23/cyberfaith-how-americans-pursue-religion-online/.

About that time a book appeared that became influential in seminary circles, titled *Being There: Culture and Formation in Two Theological Schools*.[14] In the final chapter, the authors suggested implications for seminary education that emerged from their ethnographic examination of two theological schools. Their third implication or recommendation stated, "Formative education requires prolonged and intensive exposure to an educational institution."[15] They expressed deep concerns about part-time students, extension education, and distance education, which was just emerging as a viable delivery option. They wrote, "We also have doubts about 'virtual' education. . . . Few of the new forms and technologies seem to us to deliver the full benefits of actually being there, on location at school, in its buildings, with its various populations."[16]

Every time we read this description or similar ones that convey an idealistic view of the campus community, our minds go to the vivid description Urban Whitaker provided in the earliest days of distance education. Describing the idealized community of scholars, he wrote,

> Students hurry from the bus or car (in a distant parking place) to the classroom, arriving just as the session starts, and speed away afterward (or a few minutes early in order not to be late to work). They hardly notice the other members, faculty or students, in the briefly-visited "community of scholars." The professor is often on a similar schedule, getting to campus either on Tuesday-Thursday or Monday-Wednesday-Friday to conduct a class and to meet some short, mandatory office hours (that are more than occasionally violated).[17]

One of the most frequent criticisms of online theological education is the supposed inferior status of mediated presence compared to the assumed influence of embodied presence made possible only when persons physically gather with one another. Douglas Groothuis cautions against the "ill-conceived" temptation to "educate the soul through cyberspace."[18] He added that "embodied fellowship is an irreducible and incommensurate quality

---

[14]Jackson W. Carroll, Barbara G. Wheeler, Daniel O. Aleshire, and Penny Long Marler, *Being There: Culture and Formation in Two Theological Schools* (New York: Oxford University Press, 1997).

[15]Ibid., 274.

[16]Ibid., 276.

[17]Urban Whitaker, "Inputs, Outcomes, and Nonsense About the 'Community of Scholars,'" in *Roads to the Learning Society*, ed. Lois Lamdin (Chicago: Council for Adult and Experiential Learning, 1991), 23.

[18]Douglas Groothuis, *The Soul in Cyberspace* (Grand Rapids: Baker, 1997), 150.

which cannot be adequately translated into any other form of communication."[19] We observe that the sharpest criticisms focus on the problem of presence, and no matter what technologies one employs in the educational exchange, none substitutes for the bodily presence of instructors and students. By presence, those who use the term mean "the degree to which geographically dispersed agents experience a sense of physical and/or psychological proximity through the use of particular communication technologies."[20] In this regard, Rosemary Lehman and Simone Conceicao helpfully differentiate between "cognitive presence" (thinking through re-flection and dialogue), "social presence" (personal and emotional connec-tions to the group), and "instructor presence" (the voice of the facilitator who models critical discourse, provides constructive critique, and offers formative feedback).[21]

This criticism of online community seems to contradict orthodox Christian theology regarding the communion of the saints and the com-munion believers enjoy with Christ through the Spirit. From the second century onward, the Christian church taught the doctrine of "the com-munion of the saints" (*communio sanctorum*) based upon the apostle Paul's doctrine of the spiritual union of the body of Christ. The phrase summa-rized the firm conviction that "the faithful on earth, as fellow members of the kingdom of God and members of the body of Christ, are united in bonds of deepest spiritual fellowship."[22] Karl Barth recognized the unique ministry of the Holy Spirit to this dispersed community in the world that makes pos-sible its growth when he wrote, "The Holy Spirit is the quickening power with which Jesus the Lord builds up Christianity in the world as His body, i.e., as the earthly-historical form of His own existence, causing it to grow, sustaining and ordering it as the communion of the saints."[23]

This ministry of the Holy Spirit is not only dispersed throughout the world but Barth goes on to explain that the Spirit's work extends from "the

[19]Douglas Groothuis, "Christian Scholarship and the Philosophical Analysis of Cyberspace Tech-nologies," *Journal of the Evangelical Theological Society* 41 (1998): 631-40.
[20]Esther Milne, *Letters, Postcards, Email: Technologies of Presence* (New York: Routledge, 2010), 165.
[21]Rosemary Lehman and Simone Conceicao, *Creating a Sense of Presence: How to Be There for Online Students* (San Francisco: John Wiley & Sons, 2010).
[22]J. P. Kirsch, *The Doctrine of the Communion of Saints in the Ancient Church* (Middletown, DE: Christ the King Library, 1911), 46.
[23]Karl Barth, *Church Dogmatics* IV/2, §67, 614.

right hand of God the Father" and serves as "the power which quickens from above, from a distance, from God; from the God who dwells in light unapproachable."[24] Thus, the communion of the saints exists and grows both in spite of the dispersed nature of its existence in the world and through the effective ministry of the Holy Spirit operating at a distance but unconstrained by it.

Distance is no barrier to the work of the Holy Spirit among the community of saints since their fellowship is a spiritual one not determined by physical proximity. As George Thomas Kurian put it, "In the Spirit, Christians are always present and never absent."[25] Although this spiritual fellowship may manifest itself in concrete gatherings, assemblies, and most significantly in table fellowship, these are not necessary to justify the Spirit's work. The Holy Spirit carries out his ministries among us when we are gathered and when we are scattered, when we are present with one another and when we are absent from one another. We are incapable of restricting the Spirit's various ministries to the church regardless of the form the church's gatherings reflect. The late Thomas Oden put it this way, "The redeemed are not separated from each other because they are not separated from Christ. This is a real and enduring communion of each one with Christ, and of each one with all those who share life in Christ."[26]

A characteristic of Pauline community proposed by Robert Banks is that it is "centered primarily around *fellowship*, expressed in word and deed, of the members with God and one another."[27] Since the apostle Paul used the language of *koinōnia* ("fellowship") more frequently than any other New Testament writer, it is a peculiarly Pauline word that expresses his understanding of the unique relationship that exists between the church and Christ and among Christians. This *koinōnia* community cannot be restricted to physical presence since Paul says we enjoy this fellowship with Christ now (1 Cor 1:9). Further, he insists that when the church celebrates the Lord's Supper we experience a *koinōnia* with Christ even though he is

---

[24]Ibid., 652.
[25]George Thomas Kurian, "Communion of the Saints," in *Encyclopedia of Christian Education*, ed. George Thomas Kurian and Mark A. Lamport (Lanham, MD: Rowman & Littlefield, 2015), 317.
[26]Thomas C. Oden, *Life in the Spirit* (Peabody, MA: Prince Press, 2001), 447.
[27]Robert Banks, *Paul's Idea of Community: The Early House Churches in Their Historical Setting* (Grand Rapids: Eerdmans, 1980), 111.

not physically present (1 Cor 10:16). There appears to be no indication in the New Testament that the fellowship the church enjoys with Christ now is somehow inferior to the fellowship the church experienced with him previously or will experience with him in the eschaton when we will be in his presence forever. *Koinōnia* is also used to describe our relationship to the Holy Spirit with whom we enjoy communion (2 Cor 13:14). Again, physical, face-to-face community cannot be required since we are dealing with a Spirit who enjoys no corporeal manifestation.

If we expand beyond the Pauline usage, we encounter John's unique use of the word in reference to our relationship with God (1 Jn 1:3, 6). However, no physical encounter with God is required to understand this fellowship as authentic or real. There seems to be no warrant for supposing some gradation of community from physical to spiritual. Those who were privileged to experience the real presence of Jesus or who enjoyed a direct encounter with God are never said to have a superior or even ideal experience from the rest of us who enjoy the presence of Christ and God in mediated dimensions. One could even argue that the incarnation of Jesus, God embodied in human form, is a mediated form of revelation. Would anyone want to conclude from this that, therefore, it is an inferior form of communication?

From a secular perspective, Marc Smith and Peter Kollock discuss this issue in their treatment of virtual and real-life communities. They maintain that there is much less dichotomization than we may think; they find evidence to support the notion that each serves to influence the other. They report that "many community ties connect offline as well as online. It is the relationship that is the important thing, and not the communication medium."[28] Many people cross-communicate between media. We see people sitting next to each other while texting information back and forth, and we see people separated by great distances using video-based software to chat and connect. Further, reciprocity is viewed not necessarily as an equal exchange but rather as a sort of equivalency. For example, Howard Rheingold pointed out that "the person I help may never be in a position to help me, but someone else might be."[29] The greater good is promoted

---

[28]Marc Smith and Peter Kollock, *Communities in Cyberspace* (London: Routledge, 2001), 182.
[29]Howard Rheingold, *The Virtual Community: Homesteading on the Electronic Frontier* (Cambridge, MA: MIT Press, 1993), 49.

through acts of reciprocity and interchange, which grow and sustain a community. This is further highlighted by Nicholas Christakis and James Fowler, who found that altruistic and generous behavior increases when the recipient and giver both benefit, either directly or indirectly. They discovered that "if someone was on the receiving end of a generous exchange [of money], that person would become more generous to the next set of partners."[30]

There are endless discussions about whether or to what extent community can be formed online and the degree to which it serves to influence growth, both corporately and individually. That concern is at the heart of Putnam and Feldstein's discussion about community formation over websites like Craigslist. They note that "to argue that the million-plus people who visit the San Francisco craigslist site every month form a 'community' would, we believe, reduce that word to almost meaninglessness."[31] However, in the story that unfolds about the use of Craigslist for purposes other than finding or posting a want ad, Putnam qualifies his position by noting that "craigslist has elements of community to a surprising degree and that its community nature has a great deal to do with elements that we see in other forms of community: localness, member participation . . . and purposes beyond that of simply being together."[32] One cannot ignore the rapid rise of social media and the influence this has had on many different kinds of community. If the issue were simply limited to a more convenient way to communicate with others, it is likely that popularity and even codependence would have given way to passing fads and trends. It appears that the interaction one has with members of her or his social network has enough of a powerful influence to keep that person connected. To the degree that people return time and again to the source of this influence, digital media's effects on one's formation have the potential to be positive in nature. We have observed—through reports of connectedness to other online students, exhortation to continue studying despite discouragement, and encouragement during difficult times—that online students experience growth and formation.

---

[30]Clive Thompson, "Are Your Friends Making You Fat?," *New York Times Magazine*, last modified September 10, 2009, www.nytimes.com/2009/09/13/magazine/13contagion-t.html
[31]Putnam and Feldstein, *Better Together*, 234.
[32]Ibid., 240.

In one particular course, a student shared some difficult life circumstances and the resulting emotional and mental toll they took on her. The other members of the course shared positive comments, personal notes of encouragement, and even wrote out the prayers they had prayed for that student. The student in question reported that this online group, with members of which she had never met, had embraced her in a way that her own community of physical presence had not been able to. Online community then seems to be more than simply a convenient way to learn, socialize, and engage others. This form of interaction also serves to facilitate—even *instigate*—positive formation in the lives of its members. This does not mean that every online experience, even those created in Christian communities, is positive and formative. Nor is every embodied experience of Christian community positive and formative. The apostle Paul addressed disruption of the fabric of Christian community in the church at Corinth because of social, economic, and spiritual fissures that created disunity and lead to Paul's charge of spiritual immaturity (1 Cor 3:1-4). All forms of Christian community must exercise care "to preserve the unity of the Spirit in the bond of peace" (Eph 4:3). Otherwise, the spiritual connections and interactions that form the spiritual ecology of the body of Christ get disconnected and disrupt the flow of spiritual nutrients and resources so necessary for spiritual growth individually and corporately.

CONNECTEDNESS ONLINE

There is a sense in which the attraction to growing together online reflects what Jesse Rice refers to as the need to find a sense of home or a return to place, not in the sense of location but to a place of belonging and connectedness.[33] He cites the explosive growth of Facebook as an example of this innate desire to connect with others. He noted that in just six months' time, Facebook membership, which he distinguishes from users, doubled from "one hundred million to two hundred million."[34] Rice believes connection—and the pursuit of it—lies at the heart of our fascination with Facebook. It is, in essence, part of our creational DNA and is a reflection of what God

---

[33]Jesse Rice, *The Church of Facebook: How the Hyper-Connected Are Redefining Community* (Colorado Springs: David C. Cook, 2009).
[34]Ibid., 28.

desired to have with us from the very beginning. As Stanley Grenz writes, "The transgression of our first parents led to the unmistakable disruption of community. Their act brought alienation or estrangement where once had been only fellowship."[35] The fragmented community of Babel reflects the sin of setting self above God rather than as God intended: self in relation to God and others. Rice notes that "connection is the very core of what makes us human and the very means by which we express our humanity."[36] Our original human connections with God and others were lost, and we have been trying to reconnect ever since. The issue is not that online communities have necessarily fragmented our physical spaces but that they reflect our present condition: that we yearn for connection to one another because of what we have lost and how we were created by God. We know from the Genesis account of creation that part of what it means to reflect the image of God is to connect with other humans. There is a certain community structure identified in the creation texts represented in part by plural pronouns. Genesis 1:26-28 says,

> Then God said, "Let us make mankind in our image, in our likeness, so that they may rule over the fish in the sea and the birds in the sky, over the livestock and all the wild animals, and over all the creatures that move along the ground."
>
> > So God created mankind in his own image,
> > in the image of God he created them;
> > male and female he created them.
>
> God blessed them and said to them, "Be fruitful and increase in number; fill the earth and subdue it. Rule over the fish in the sea and the birds in the sky and over every living creature that moves on the ground." (NIV)

The Genesis 5:1-2 text reminds us of a similar plurality as a representation of God's likeness; it maintains that "when God created mankind, he made them in the likeness of God. He created them male and female and blessed them. And he named them 'Mankind' when they were created" (NIV). The image of God we bear enables or predisposes us to seek connections with others that reflect the interconnections among the members of the Godhead.

---

[35]Stanley J. Grenz, *Theology for the Community of God* (Grand Rapids: Eerdmans, 1994), 188.
[36]Rice, *Church of Facebook*, 28.

We are well aware that online community is not always a positive contributor to growth and formation, but communities in physical spaces don't always contribute to the growth of the individual either. We factor in this reality as we consider ways of facilitating community, both online and off, that reflect environments conducive to growth. The development of online community has been shaped in large part by how we define and use Internet-based technologies. Rheingold noted, "The technology that makes virtual communities possible has the potential to bring enormous leverage to ordinary citizens at relatively little cost—intellectual leverage, social leverage, commercial leverage, and most important, political leverage."[37] People naturally yearn for and seek out community, which may explain why the Internet has become such a positive and popular contribution to online social development. Rheingold defined virtual communities as groups that form as a result of online interaction over a period of time and with an emotional investment. Much of what happens in a physical community can be replicated in a virtual community. Rheingold observed, "People in virtual communities do just about everything people do in real life, but we leave our bodies behind."[38] However, Mary Hess has argued that online education does not, in fact, lead to a disembodied nature of learning.[39] The individual is very much interacting with the medium of technology through the use of the mental, physical, emotional, and spiritual dimensions. We do not simply grow alone; we grow together in relation to others.

Online experiences show us that we can give and receive care for one another, value interpersonal relationships, and interact with one another, thereby creating connections that form community. Heidi Campbell highlights this in her descriptions of self-reported attributes of online community. She describes six components that members in her study reported about religious online community development. In particular, those attributes included experiences related to relationship development. Some participants even reported that "they experienced close relationships with

---

[37]Rheingold, *Virtual Community*, xix.
[38]Ibid., xvii.
[39]Mary Hess, "Attending to Embodiedness in Online, Theologically-Focused Learning" (paper written for Colloquium "Going the Distance: Theology, Religious Education, and Interactive Distance Education," University of Dayton, November 1999), http://meh.religioused.org/dayton.pdf.

others online."[40] Care was another attribute that online community members experienced through the ability to "give and receive support and encouragement."[41] Through this experience, members felt that they were heard, validated, and accepted in ways they did not experience in offline communities.

Feeling valued was another attribute of online communities, underscored by the idea that "many online communities have a long-term, established membership where people are consistently encouraged to stay committed and involved."[42] People yearn to belong to something of value and significance. Connection was an attribute of her study that emanated from communication and access to participating members. Campbell noted that connection was a way to bridge the gap between physical gatherings. We often borrow the rail fence analogy to illustrate this same concept.[43] The posts of the fence are regular gatherings, such as weekly or monthly class sessions, while the rails represent online connections we make with one another throughout the duration of the course. Intimate communication was yet another attribute that online community members reported in their experiences and was described as "a safe place for transparency and encouraged accountability."[44] Further, online community seems to encourage and allow for members to "go deeper more quickly" and bypasses the small talk that many believe is "not essential for establishing relationships and interaction."[45] This level of intimacy generates reciprocity in that relationships formed and fostered offline are deepened through online communication and "intimate communication online also increases individuals' desire to have face-to-face contact, to extend their relationship to a new level."[46]

---

[40]Heidi Campbell, *Exploring Religious Community Online: We Are One in the Network* (New York: Lang, 2005), 181.

[41]Ibid., 182.

[42]Ibid., 183.

[43]Ted W. Ward and Samuel F. Rowen, "The Significance of the Extension Seminary," *Evangelical Missions Quarterly* 9, no. 2 (1972): 17-27. Although many use this illustration as their own, it originated with Ted Ward and Sam Rowen while Ted was professor of education and Sam a doctoral student at Michigan State University. Ted originally conceptualized theological education by extension, using the rails to refer to cognitive inputs from readings and other resources and the fence posts represented the in-person seminars or class sessions held periodically.

[44]Campbell, *Exploring Religious Community Online*, 185.

[45]Ibid.

[46]Ibid.

Finally, Campbell pointed out that online religious communities form around a common or collective core of faith, which differentiates it from other online groups. She noted that "this shared faith also provides a common language, which supports members' identification and investment in one another."[47] In all of this, we see a common element of reciprocal growth, instigated by the coming together and connection with one another online.

ONLINE CONTEXTS

Students who engage online experience an environment that shapes and forms their development. These students bring with them a vast network of social connections that influence what and how they learn. Students also contribute a wealth of knowledge, experience, and the ability to connect personally with others. Furthermore, students who read course texts are interacting with the words, thoughts, and expressions of the representative contributors. When they interact with others online, they engage with another person's network, which serves to inform and influence and, ultimately, leads to continued growth.

Some would argue that online activity by its very nature creates and contributes to isolation and disconnection from others. The reality, however, is that isolation and disconnectedness have become a part of American life over the last several decades, even prior to the rise of the Internet and online social network platforms. Putnam and Feldstein wrote, "Beginning, roughly speaking, in the late 1960s, Americans in massive numbers began to join less, trust less, give less, vote less, and schmooze less."[48] Ray Oldenburg lamented a loss of an informal public place, similar to coffee houses in Europe, where people could gather for social discourse.[49] He maintained that the problem of place is magnified by the inability to establish common experiences outside of family or professional contacts. Oldenburg's solution for the problem of place is what he termed "the third place," components in society that allow for the engagement of sociopolitical structures. The third place

---

[47]Ibid., 186.
[48]Putnam and Feldstein, *Better Together*, 4.
[49]Ray Oldenburg, *The Great Good Place: Cafés, Coffee Shops, Bookstores, Bars, Hair Salons and Other Hangouts at the Heart of a Community* (New York: Marlowe, 1989).

has both tangible and intangible evidence; places such as public parks and city squares facilitate informal gatherings of people. The third place also speaks to the intangible by allowing for a level playing field so that diverse groups can come together in discourse and greater unity.

Some scholars have not embraced virtual community as a replacement for physical community. One such author, Douglas Groothuis, noted in an interview, "But overall cyberspace (and hardly anyone calls it this any more) has diminished community if one means by that embodied relationships bound by troth, friendship, citizenship, and physical proximity." He goes on to note in this interview that "with the rise of social networking—Facebook, MySpace, Twitter, etc.—the temptation to avoid the face-to-face world has increased."[50] In his book *The Soul in Cyberspace,* Groothuis takes to task the role of emerging technologies and cyberspace on the spiritual deprivation of those overly engaged in virtual communities (in contrast to those in physical communities like his classroom). In another case, Yuang Han Kim noted the apparent dichotomy in the observation "cyber-worship and churches have begun replacing traditional Christian worship and churches. This increasing phenomenon will result in a certain wearing away of the historical institutional churches and worship."[51] The concern we and others have is that the lines of demarcation cannot continue to be so closely drawn.

We add our voices to this discussion by noting that while much of the argument centers on an either/or proposition, there are benefits to be gained from both virtual and physical communities. As such, we seek to find ways to make both venues contribute to the growth and development of participants. There are mutually beneficial components to physical gatherings, but John Gresham argued that these benefits are not exclusively reserved for face-to-face gatherings.[52] John Horrigan used the term "virtual third places" to suggest that the online community has become prolific with users seeking to gain access to something more than local community. He concluded that online groups serve to enhance community rather than to destroy it:

---

[50]Douglas Groothuis, interview by Tim Challies, *Challies* (blog) May 6, 2009, www.challies.com/interviews/the-soul-in-cyberspace-an-interview-with-douglas-groothuis/.

[51]Yung Han Kim, "The Identity of Reformed Theology and Its Ecumenicity in the Twenty-First Century: Reformed Theology as Transformational Cultural Theology," in *Reformed Theology: Identity and Ecumenicity,* ed. Wallace M. Alston and Michael Welker (Grand Rapids: Eerdmans, 2003), 3-19.

[52]John Gresham, "The Divine Pedagogy as a Model for Online Education," *Teaching Theology & Religion* 9 (2006): 24-28.

In some ways, online communities have become *virtual third places* for people because they are different places from home and work. These places allow people either to hang out with others or more actively engage with professional associations, hobby groups, religious organizations, or sports leagues.[53]

Online communities described. Palloff and Pratt use the terms *community* and *communicate* to suggest a semantic connection in which people seek to come together through the process of the shared word.[54] Putnam reiterated this point by noting that "community, communion, and communication are intimately as well as etymologically related."[55] Palloff and Pratt similarly argued, "Our attempts to communicate are attempts at community building."[56] Technological advancements have shaped our ability to communicate, and the growing body of resources found in pedagogical media are forcing some of us to rethink previously held notions of limited communication. Online communication may, in fact, lead to deeper dialogue. Campbell's research led her to conclude that for many, local relationships did not necessarily yield the in-depth conversations, honesty, and transparency members sought but "online it can be easier to find others who share commonality of experiences and belief and to discuss these in an open forum."[57] Gresham stated, "Within the physical classroom, social and environmental conditions conspire to limit the full involvement of all students in a class discussion."[58] One of the dynamics we have observed in the physical classroom is how social barriers, perceived or otherwise, contribute to lack of participation. Mary taught a seminary course in which a student reported social awkwardness and the desire not to "look stupid" in front of others. He would rarely contribute to classroom discussion. That same student took a different course with Mary online, and she could hardly keep him contained. He wrote profusely and other students

---

[53]John B. Horrigan, "Online Communities: Networks That Nurture Long-Distance Relationships and Local Ties," Pew Internet and American Life Project, October 31, 2001, www.pewinternet .org/files/old-media//Files/Reports/2001/PIP_Communities_Report.pdf.pdf.

[54]Rena Palloff and Keith Pratt, *Building Learning Communities in Cyberspace* (San Francisco: Jossey-Bass, 1999).

[55]Robert Putnam, *Bowling Alone: The Collapse and Revival of American Community* (New York: Simon & Schuster, 2000), 171.

[56]Palloff and Pratt, *Building Learning Communities*, 25.

[57]Campbell, *Exploring Religious Community Online*, 185.

[58]Gresham, "Divine Pedagogy as a Model," 27.

vigorously interacted with him and his ideas. We've seen other examples in which students hesitate to start a discussion in class because they're concerned the conversation will spill over outside of class when they're in a hurry to leave. In some cases, the campus environment no longer contributes to expanded in-person dialogue because of the realities of commuter life. The underlying value in communication as a way of building community is its ability to unify or bring together individual components into a cohesive whole.

The notion of community is not static or isolated. Jennifer Lock suggests that community can be compared to a developmental process that mirrors face-to-face communities. Just as physical communities do not remain unchanged, learning communities work in tandem with the same dynamics to deepen and solidify relationships. She suggests that in order for an online community to develop, it must receive the same attention from its members as it does in face-to-face communities. The formation of community is largely dependent on the reciprocity of relationships: "The relationships, the intimacy, the negotiations, and the engagement of participants all influence the evolution of a community. The growth and longevity of a community are directly related to the community meeting the needs of members."[59] Online socialization does not necessarily preclude intimacy, according to Smith and Kollock. There is a relationship between virtual dialogue and feelings of closeness between correspondents. They concluded that "people on the Internet have a greater tendency to base their feelings of closeness on the basis of shared interests rather than on the basis of shared social characteristics such as gender and socio-economic status."[60]

Formational contexts. The formational development faith-based institutions seek to provide is as much a part of community formation as any other area of development. In fact, Palloff and Pratt use Mezirow's term "transformative learning" to highlight not just "real learning" but to suggest that through this process of self-reflection and engagement with others, complete personal transformation occurs.[61] Formation and transformation are

---

[59]Jennifer V. Lock, "Laying the Groundwork for the Development of Learning Communities Within Online Courses," *Quarterly Review of Distance Education* 3 (2002): 395-408.

[60]Smith and Kollock, *Communities in Cyberspace*, 186.

[61]Palloff and Pratt, *Building Learning Communities*, 129.

not limited strictly to the spiritual dimension; they include the entire eco-system of a person. Additionally, this formation expands from personal transformation to one that involves and necessarily includes a person's net-works. Online students, contrary to some opinions, are not necessarily au-tonomous learners; rather, they represent a cohort of connections, in-cluding family, ministry, neighborhoods, and friends. Palloff and Pratt, in describing the transformative learning process, use terms like "acquiring knowledge," "form new relationships," and "acknowledgement of ideas."[62] They describe the growth process that naturally occurs in the process of learning together. There is a bidirectional relationship in the community of learning; growth occurs both individually and corporately. The learner herself forms relationships with other learners and, in that formation, is encouraged to expand her learning experience. Palloff and Pratt note that "this creates a web of learning through which new ideas and means of re-flection provide a feedback mechanism."[63] While the context for Palloff and Pratt's comments is limited to the online learning community, an eco-systems view of growth takes into account any number of settings in which people may find themselves.[64] We expect formation to occur online just as in a face-to-face encounter.

Spiritual formation online. Growth and development should maintain a healthy tension between intentionality and naturally occurring patterns of growth. Elizabeth Patterson argued, "We cannot assume that formation is happening automatically simply because a classroom does exist."[65] Similarly, we cannot assume that formation is happening simply because we teach a course on spiritual disciplines in an online classroom. Spiritual formation is not a magical occurrence resulting from the presence of Christians gathered together in the same place, whether online or on campus. There must also be intentionality, reflection, engagement, and interaction between and among those gathered.

---

[62]Ibid., 129.

[63]Ibid., 131.

[64]Ivan Illich, *Deschooling Society* (New York: Harper & Row, 1970), proposed the concept of "learning webs" that could be created to connect various sources of knowledge and information into a network of learning resources that made learning more distributed, student centered, and flexible.

[65]Elizabeth Patterson, "The Questions of Distance Education," *Theological Education* 33, no. 1 (1996): 59-74.

One example of intentional interaction is through blogging. David Wayne cites Tim Bednar's support for blogging as a method of building religious community, claiming that it allows people to take a more active role in their personal spiritual formation.[66] Bednar argues that blogging encourages a participatory form of Christianity and has a salutary effect on spiritual formation. Blogging for religious purposes differs from other religious webpages in that it promotes an interactive and proactive form of communication. Furthermore, it encourages the user to assume ownership for the posted content. Those who encounter religious portals, in contrast to blogging, are unable to take the steps necessary to develop formative changes or responsibility for their spiritual formation. We don't seek out blogs as a way to replace spiritual leadership and influence but rather to intentionally engage in our own growth and formational processes. Moll noted, "Internet seekers remain connected to their local church, but they pursue their own spiritual interests online. Blogging is an attractive spiritual discipline precisely because it is unmediated by our church or pastor."[67]

SUMMARY

People are coming together online in ever-increasing numbers motivated by a desire to grow spiritually. Whether or not those who do so are actually growing spiritually is debated within the church and Christian academic institutions. Our "ecologies of faith" model of spiritual formation offers an understanding of faith formation in authentic online communities that recognizes not just a virtual reality but a spiritual one. In the next chapter, we will examine ways to engage others through various means of creating and designing online learning ecologies that foster genuine faith.

---

[66]David Wayne, "Tim Bednar's Paper: We Know More Than Our Pastors," *Jolly Blogger*, July 1, 2005, http://jollyblogger.typepad.com/jollyblogger/2005/07/tim_bednars_pap.html#more.

[67]Rob Moll, "Blogger Predicts Revival via Web," *Christianity Today*, Fall 2004, www.christianity today.com/pastors/2004/fall/4.13.html.

— SIX —

DESIGNING LEARNING ECOLOGIES

A learning ecology requires a unifying
model of instructional theory that, in turn, drives
the architecture of such a learning environment.

ARTHUR RICHARDSON, "AN ECOLOGY OF LEARNING
AND THE ROLE OF ELEARNING IN THE
LEARNING ENVIRONMENT"

During the harrowing days of the Battle of Britain in World War II, an American flying for the Royal Canadian Air Force conducted a test flight of the latest version of the British Spitfire fighter plane. As the pilot soared above the earth at thirty thousand feet, he had an epiphany and began composing in his head the words to a poem that he finished when he returned to his barracks. John Gillespie Magee Jr. titled his poem "High Flight" and opened with these familiar word: "Oh! I have slipped the surly bonds of earth. . . . I've trod / the high untrespassed sanctity of space, / Put out my hand, and touched the face of God." His high-altitude view gave him a perspective and insight he could never have achieved while inhabiting the "surly bonds of earth." We think that an ecology of learning gives us just that sort of high-altitude perspective. It allows us to assume a meta-perspective that embraces and encompasses the totality of the learning environment. When approaching a task like designing an online learning environment, we must zoom out before we zoom in. In this chapter we want to take a wide-angle look at the online learning ecology and recognize the various components that conspire to facilitate learning and understanding.

A learning ecology, then, is "the space in which learning occurs that operates on the same principles and processes as a natural ecology."[1] The space referenced here can be the *physical* space of an on-campus classroom or the *virtual* space of an online course. It may even be the *personal* space of an individual student who creates a unique learning ecology comprising what Norman Jackson calls "lifewide experiences."[2]

The primary feature of any ecosystem—whether natural, social, spiritual, or educational—is that it is interconnective, interactive, dynamic, and mutualistic in terms of the shared benefits that pass between individual organisms. All organisms need an ecosystem, including student organisms in an educational setting.

As we discuss the design of learning ecologies in this chapter, it is important to remember not to allow "the tail to wag the dog." It is easy for those of us living in a high-tech society to become easily enamored and hypnotized by technological gadgetry and forget about the human agents involved. The centerpiece of God's creation has not been supplanted even though our ancient enemy might prefer that. The focus of any learning ecology remains on the students, professors, course designers, and other human contributors who make it effective. While we will necessarily address matters of technology and infrastructure, we must not lose sight of the most important part of the learning ecology: our students and the professors who provide guidance and instruction.

In a course Mary taught for United Theological Seminary, she set up a private Facebook group for the students enrolled in that section. Although students were required to participate, they demonstrated contributions that went far above the minimum expectations. In one example, a student asked for prayer for her trip abroad. The other learners responded by posting their prayers, offering to pray each day, giving advice for her travels, and asking that she check in from time to time. She did that, both with written updates and with photos. In another post, one student shared pictures of her grandchild just as a way of telling us about her family.

---

[1] G. Siemens, "Collectivism: Creating a Learning Ecology in Distributed Environments," in *Didactics of Microlearning: Concepts, Discourses, and Examples*, ed. T. Hug (Munster, Germany: Waxmann Verlag, 2007), 63.

[2] Norman J. Jackson, *Exploring Learning Ecologies* (London: Chalk Mountain, 2016), 4-5.

Another student would regularly share articles he found that related to the course content, whatever the subject was at the time. Interestingly, he continued to send articles long after that class ended. We used the Facebook group to determine the best times to meet for synchronous sessions, post and get clarification about course assignments, and solicit assistance with professional development. One student asked for advice about how to set up a particular learning strategy for his own course and not surprisingly, he continues to communicate with various members of the class via email. Another student came back to the Facebook group six months after the course ended, stating, "I came back to this page for ideas as I am creating my spring online course weekly schedule. Looking forward to incorporating different social media tools." Finally, one participant in this Facebook group contributed a significant number of teaching strategies to her fellow classmates. It just so happened that this was a particular topic of interest for a conference Mary directed a couple years later and she was able to shoulder-tap this individual for the keynote address at the conference. In all of this, the Facebook group members interacted with one another, prayed for each other, supported and encouraged one another, and created some connections and interactions that went beyond simply posting course content.

In its earliest form, many viewed digital education as a compartmentalized orientation to teaching and learning, but as education has evolved, we have come to understand that teaching and learning online are much more expansive and interconnected than previously assumed. Instructional manipulations or modifications in one area of the teaching and learning process affect other areas, whether this is intended or not. Walter Dick et al. note that "changes in one component can affect other components and the eventual learning outcomes."[3] Their conclusion is that failure to account for this interactive character of online learning "can doom the entire instructional process."[4] Digital learning needs to take a systems approach to education. By "system" Dick and his colleagues mean "a set of interrelated parts, all of which work together toward a defined goal."[5] We come at this issue in much the

---

[3] Walter Dick, Lou Carey, and James O. Carey, *The Systematic Design of Instruction*, 8th ed. (London: Pearson, 2014), 1.
[4] Ibid.
[5] Ibid.

same way, except we use the language of ecology, which is another form of a system that operates in nature. A systems or ecological approach to the design of online learning takes a global or comprehensive view of the teaching-learning process, and this view includes an expectation that students have valuable knowledge, experience, or perspective to share. Students who study online bring to the experience a network of embedded connections, and we think an online course designer should consider those social connections and relationships to help construct a meaningful online learning experience.

The relationships students have with members of their church, individuals within their family, or people in the workplace can influence what takes place in an online course. Conversely, the classroom content and interactions in the academy can often influence other people within an ecosystem. A student in the classroom who is caring for an aging parent is likely to contribute more significantly to an online class about older adults because the student has an immediate, personal experience to relate to the course content. This contribution not only affects the student's understanding of how to minister to older adults but it helps him or her connect emotionally to fellow students and informs the class's prayer support of that classmate. This isn't to say that others can't contribute significantly without that personal experience, but we more fully embrace the issue of contexts or ecosystems when we incorporate those experiences as part of the learning process.

We understand that the ecosystem of a student's range of social connections contributes to how they grow and develop spiritually, and as such, we embrace various interactions as part of the learning process. While it is important to be intentional about growth and development, we are not convinced that to be successful every course we teach must specifically address all the ways in which students can grow. Social development, for example, may be occurring online, in the on-campus classroom, or in the Sunday school class. Put another way, bidirectional influences work to affect both the student and his or her sources of influence. Dick et al. similarly discuss their systems approach as a way of orienting their readers to understanding the various components of instruction and the need for these components to "interact in order to achieve the goal."[6] Rena M. Palloff and Keith Pratt

---

[6]Ibid., 2.

touch on this notion by highlighting the interrelatedness of their proposed system of learning design.[7]

Effective instruction requires both content knowledge and interpersonal skill. We know that instructors' primary challenges are those involving human interaction.[8] In our drive to employ the latest technologies, we occasionally forget about the human touch, but digital or technological devices can nonetheless serve to enhance or contribute to basic human interaction in a way that is comparable to face-to-face encounters. The learning interactions and transactions that occur in online courses are simply another form of human interaction and communication. The human element is never missing because humans are employing the technology to engage one another. As long as the instructor employs some type of exchange among participants, no matter what form it takes, a course has the potential of educational benefit to the students and faculty involved in the exchange. While community building is always part of the teaching-learning process, intentionality in course design plays a critical role in successful human interaction—for example, in planned reciprocal exchanges between individuals within a class.

CONTEXTUALIZED LEARNING

Designing learning ecologies requires a certain proficiency with systems and strategies that can enhance, sustain, and facilitate that system. Regardless of the learning strategies used, reciprocal engagement is at the core of what it means to nurture the spiritual. Developing and designing a learning ecology requires the realization that technology alone cannot create or maintain human relationships and should not attempt to replace them. If institutions try to find technological ways to replace or remove face-to-face encounters, there is reason for concern.[9] Instead they should focus on technology that encourages individuals to come together in community or to contribute to the external contexts within which a student operates. Steve Delamarter and Dan Brunner maintain that the old dichotomies within many online or

---

[7]Rena M. Palloff and Keith Pratt, *Lessons from the Cyberspace Classroom* (San Francisco: Jossey-Bass, 2001).

[8]Ken White, "Face to Face in the Online Classroom," in *The Online Teaching Guide*, ed. Bob Weight and Ken White (Needham Heights, MA: Allyn and Bacon, 2000), 1-12.

[9]Scott Cormode, "Using Computers in Theological Education: Rules of Thumb," *Theological Education* 36, no. 1 (1999): 101-16.

distance education programs are giving way to models of teaching and learning that involve a broader view of facilitating information.[10] The paradigm is shifting from an either/or model to one that embraces various contexts and ecosystems. Seminary education that occurs online allows students to remain in ministry contexts while simultaneously furthering their education. This allows a bidirectional, interactive experience that speaks to multiple systems—including the ministry context—of participating students.

Delamarter and Brunner observed that when teaching and learning focuses on technology, many programs face termination due to unrealistic expectations about the work, time, and money involved to sustain them. They suggested, however, that when we view teaching and learning from a pedagogical or andragogic perspective, the process is successful because the issues are centered on learning strategies that allow for both online and physical interaction. Delamarter and Brunner concluded from their research that "educators claim that both they and their students have been surprised by the depth of community and the vibrancy of learning that take place in the online environment."[11] Ken Bedell writes, "Electronic media provide the environment (just as books and printed material did before) where people experience communities of meaning, form their own identity, and mature spiritually."[12]

It seems that we have to begin or continue engaging our students in ways that reflect where they are in real life. Their real-life ecologies provide formative opportunities for growth. John Palka's work highlights one's ecology as a significant context for spiritual formation.[13] In this particular study, he looked at reports from seminary students about the priority of their contexts for formational input. He cautioned that while institutions may believe they are the main impetus for a student's growth and development, it is actually the community in which the student lives and works in every day real life that contributes most significantly to overall formation. In terms of those

---

[10]Steve Delamarter and Dan Brunner, "Theological Education and Hybrid Models of Distance Learning," *Theological Education* 40, no. 2 (2004): 145-64.

[11]Ibid., 147.

[12]Ken Bedell, "Technology and Theological Education" (paper presented at the School of Theology and Ministry, Seattle University, 1999), www.religion-research.org/Education.htm.

[13]John Palka, "Defining a Theological Education Community," *International Review of Research in Open and Distance Learning* 5, no. 3 (2004), www.irrodl.org/index.php/irrodl/article/view/197/279.

activities that have influence on spiritually formative experiences, learning activities (the assumption being made here is the classroom) rank only third in terms of importance. In other words, it's not just the academy that has a spiritually formative influence on students; social contexts like conversations, interaction with others, worshiping together, and living out real life with community members play just as great a part in students' spiritual formation. According to Palka's findings, the primary influence on a student's spiritual formation was the student's local congregation.

Palka found that students see social support of their spiritual formation taking place predominantly *outside* the academic community. Conversely, students perceive activities associated with the transfer of information and the construction of knowledge as primarily occurring *inside* the institution. In other words, multiple contexts influence a student's formation. The role of educators is to take advantage of those naturally forming experiences and embrace them as legitimate aspects of the student's learning ecology. We can use instructional strategies to integrate elements of the student's ecology to encourage reciprocal interaction, immediate engagement, and deepen learning and reflection.

COLLABORATION IN LEARNING CONTEXTS

One of the changes to the educational landscape has been the growth and application of collaborative learning models found, in part, with social media sites. There is a growing sense in which our students are essentially living out their life online. Recently, we watched an interview with a young girl kidnapped by a family friend and later rescued. After she was safe, she did what nearly every other teenager does—she went online and commented about her ordeal. When asked about the wisdom or healthiness of doing that, she said, "it's what we do." That's how her generation debriefs and seeks support. They go to their online network, where they find emotional support and embrace. They engage in relationships, seek advice, and share deeply on social networking sites. They feel more comfortable in an environment that provides them anonymity, anytime access, and digital convenience. In fact, nearly 60% of teenage social networkers said they look to their online social networks for advice.[14] We know from the work of Boyd

---

[14]Nielsen Reports, "How Teens Use Media," October 7, 2010, www.nielsen.com/us/en/reports/2009/How-Teens-Use-Media.html.

and others that social media are not the salve that soothes all wounds.[15] They can promote an atmosphere of distrust, social inequality, and economic divides. As with spiritual formation in general and online course design in particular, leveraging social media and networking tools to promote healthy and informative uses requires intentionality.

It isn't just teens and young adults who are living out their lives online. Investigators at the Pew Research Center for Internet and Technology report an uptick in the usage of social media across all platforms. Nearly 70% of all Americans are Facebook users while "28% use Instagram, 26% use Pinterest, 25% use LinkedIn and 21% use Twitter."[16] Roughly 60% of all Americans get their news from various websites with approximately half of those sourcing social media for this information.[17] While there are any number of negative uses found in social media (much like you would likely find in any medium), Barna noted that the primary purpose or function of social media is connection. We go online to connect with others. Barna's research supports this by showing that "54% of parents feel more connected to friends and family," while a majority of women (81%) "report feeling connected to friends through social media."[18]

One of the more interesting findings in our research revealed that social media are contributing to a shift from a more individualistic approach to one that focuses attention on collaboration with others. As we'll discuss in the next section, online collaboration is something that young adults somehow inherently understand and can leverage to serve their own needs. Much of the preference for collaborative learning is facilitated by the use of social networking sites and media tools. Karen Kear makes note of the fact that there is much more buy-in by educators with web-based learning and other tools found in social media. Mary experienced this with a presentation by a colleague regarding the use of Twitter to complete an assignment for a sociology class. The students were required to travel throughout the city's urban areas using public transportation. They were given various categories

---

[15]Danah Boyd, *It's Complicated: The Social Lives of Networked Teens* (New Haven, CT: Yale University Press, 2014).

[16]Shannon Greenwood, Andrew Perrin, and Maeve Duggan, "Social Media Update 2016," Pew Research Center Internet & Technology, November 11, 2016, www.pewinternet.org/2016/11/11/social-media-update-2016.

[17]Barna Group, *Barna Trends 2016* (Grand Rapids: Baker Books, 2016).

[18]Ibid., 28.

of locations to visit and then tweet back to the class what they had found. One particularly poignant example came from a student who tweeted she had never ridden a city bus and had no idea she needed cash to get around. Another student tweeted pictures of homeless persons and her interactions with them. Kear noted that "by using online networking tools such as discussion forums, social network sites, wikis, blogs, and instant messaging, learners can carry out activities together."[19] Kear goes on to qualify her observation by recognizing that simply utilizing these tools doesn't necessarily create a community of learning. Community happens when there is a certain degree of intentionality toward collaborative learning facilitated by those in that community. When we value relationships over technology, we are then able to implement teaching strategies that elevate the human connection over digital devices. The latter settles into its proper place in the digital economy's taxonomy as a means to a method rather than a method itself.

One well-placed concern about creating digital learning opportunities is that we will isolate learners from contexts that seek to educate and inform. Critics falsely assume that online learning inherently creates autonomous learners who seek to distance themselves from others. The issue, when seen from a broader perspective, is much more integrated and holistic. Learners bring their own social ecologies to the online learning experience. We often liken this to the checked baggage we carry with us on a plane. Our seatmate only sees what we carry on, but there may be several suitcases full of personal possessions underneath in the cargo hold.

The bidirectional nature of learning allows for a reciprocal exchange of ideas, knowledge, and experiences. As noted in the previous section, digital learning technologies and strategies are moving toward a much more collaborative approach to instruction that draws in various elements of the learner's ecology. Kear highlighted the dichotomization of learning when she observed that "studying is no longer an individual activity, with hours spent alone, reading books or lecture notes. Instead, there is an increased focus on collaborative learning, with students working in teams, sharing ideas and resources, and creating things together."[20] A team at

---

[19]Karen Kear, *Online and Social Networking Communities: A Best Practice Guide for Educators* (New York: Routledge, 2011), 2.
[20]Ibid., 1.

Duke University conducted an experiment to test the theory that collaborative models of learning both affect and are affected by a generation that has grown up largely attached to digital devices. The assumption they explored was that digital attachment automatically suggests detachment from social connections and collaboration. What Cathy Davidson and her team discovered, before the term became part of our vocabulary, was that university students intrinsically understood "crowdsourcing." This echoes what we know about digital natives: their reliance on technology has produced an egalitarian approach to problem solving. Davidson's experiment encouraged students to work together to develop learning apps for the first generation of iPods. Davidson noted that "interconnection was the part the students grasped before any of us did. Students who had grown up connected digitally gravitated to ways that the iPod could be used for collective learning. They turned iPods into social media and networked their learning in ways we did not anticipate."[21] Davidson noted that the term "cognitive surplus" is another way in which those in digital ecologies apply the "'more than the sum of the parts' form of collaborative thinking that happens when groups think together online."[22] She employed an approach she called "collaboration by difference" as a way to embrace the notion that we can overcome attention blindness by working together to see what we might individually overlook.

Davidson tested this approach in a class that combined residential with online components, which required students to post entries on Wikipedia, contribute to its public discourse, blog weekly, and participate in other collaborative exercises. Davidson found a somewhat curious relationship between the writing quality of online postings through blogs or wikis and the traditional research paper. She observed,

> When I had both samples in front of me, I discovered something curious. Their writing online, at least in their blogs, was incomparably better than in the traditional papers. In fact, given all the tripe one hears from pundits about how the Internet dumbs our kids down, I was shocked that elegant bloggers often

---

[21]Cathy N. Davidson, "Collaborative Learning for the Digital Age," *The Chronicle of Higher Education*, last modified August 26, 2011, http://chronicle.com/article/Collaborative-Learning-for-the-/128789/.

[22]Ibid.

turned out to be the clunkiest and most pretentious of research-paper writers. Term papers rolled in that were shot through with jargon, stilted diction, poor word choice, rambling thoughts, and even pretentious grammatical errors.[23]

A major presupposition for these bloggers is their collective approach to writing. Davidson noted that people of every age demographic take their writing more seriously when peers or the public evaluate it. There is intentional reciprocity in this community approach to writing and blogging. Her main takeaway from this experiment was that students enjoy sharing what they are learning and what they know with other people. Their Instagram and Facebook pages are full of selfies, six-second videos, and seemingly mind-numbing details of their personal lives, but that is how they connect and stay connected to one another. We must begin the process of unlearning what we have traditionally held sacred to the art of teaching and learning and accommodate newer ways of educational crowdsourcing to accomplish what we say we want.

Many of the emergent technologies identified below are largely collaborative in nature. Designing learning ecologies with this paradigm in mind creates opportunities for students to expand their reach through reciprocity and collaborative learning. The shift that we have seen in education seems to be moving in favor of broader dissemination, virtual community building for the purpose of sharing information, and releasing courses (in some cases for free) to the broader public.

DIGITAL LEARNING STRATEGIES

Faced with the range of options for creating digital learning ecologies through the use of various tools, online users can sometimes feel overwhelmed by or unsure of which options best reflect intended learning outcomes. Given overarching principles of creating environments that allow people to share experiences and facilitating dialogue that leads to formative opportunities, we have used some of the following strategies and see them as representative and illustrative of how technology can serve higher goals.

Facebook. Designing opportunities to engage a student's ecology in the learning process requires using different tools and strategies to broaden the

---

[23]Ibid.

scope of the diversity in experience and acquisition of content. People who connect to others on social media often express care for one another, make offers of goodwill, and generate mutually beneficial expressions of kindness. In one particular instance, Mary set up a private Facebook group for a course she taught online. The Facebook group was meant to allow her to observe group dynamics, frequency of use, and depth of interaction between the members of the course. She found that the students who were all digital immigrants—loosely defined as those who migrated to virtual platforms later than Millennials—managed to embrace this form of reciprocal inter- action in a way that was surprising and refreshing. Many of the students shared details of their personal lives, which in some cases allowed class members to pray in a more intelligent way, share helpful advice, and offer other tools that other students could directly and immediately apply to min- istry and life. One participant shared a real-time prayer request for one of her own students, to which others in group replied. Their private Facebook group page was flooded with comments of concern, prayers, helpful advice, and suggestions for helping others get through such a difficult time. Another course participant talked about an overseas trip and some of her concerns, fears, and trepidations about international travel and teaching. Her class- mates wrote out prayers for her, offered encouragement, shared their own travel experiences, and promised to pray for her each day of her trip. Each member of the group agreed to claim one day of her trip to pray for her.

Other Facebook activities included sharing articles related to online teaching strategies, posting links that were relevant to common experiences, rejoicing over the birth of a grandchild, and collectively expressing empathy for the emotional struggles of some who were facing personal difficulties. The class learned valuable course content from one another and from course assignments. Members brought to that experience their own networks of relationships, experiences, values, and insights so that in the process of de- veloping that particular online ecology, members were able to mutually engage one another and bring back to their own networks what they learned and experienced together.

Twitter. Another way Mary attempted to design and create learning ecol- ogies in that online class was through Twitter. While this wasn't as robust a social ecology tool for that particular group, it did serve to engage others, especially those outside the class—outside that particular concentric circle of the ecosystem. One course assignment asked students to read an article and then re-create the scope of that information in their own communities. Mary asked the students to attend a church that was different from their primary traditions and to tweet their observations about that experience. She discovered that this exercise allowed students to engage a virtual community through their own Twitter network, and members of those individual networks could be brought in on the experience, albeit vicariously. Further, it allowed students to be present in their own community networks and contribute to a collective experience.

One of Steve's students shared an experience he had with his daughter and her use of Twitter during a church service. During his sermon, he noticed that his daughter and her friend were on their digital devices and engaged in what seemed like a distracting use of social media. After the service, he asked them both to step into his office so he could gently scold them for what seemed to be inappropriate behavior during the worship service. He suggested that they pay better attention and be less distracting. What his daughter shared next was completely unexpected and moving. She said, "Dad, we were actually listening very carefully, and when you talked about sharing the gospel, we tweeted that out to our friends, just like you told us to do. We wanted to tell others about Jesus just like you were doing in your sermon." These two young women were digital evangelists engaging their social ecologies with the gospel.

Spreaker. Another other tool Mary used was Spreaker, an online application for creating and sharing live audio content across the Internet. She asked students to record thoughts, feelings, observations, and opinions about the church visit experience. They posted their observations to the online classroom and were invited to share the platform with their personal networks. One student used the Spreaker tool to report from Africa, thereby widening the scope of the students' exposure to varied church experiences.

Wikispaces. One teaching device we used was Wikispaces, which allows the course participants to contribute to an existing article on the course wiki site. The assignment was to pick a topic on that posted article and conduct some brief research on it. The students were then instructed to write a synopsis and add it to the class article. The result of this assignment yielded several things: First, it allowed each individual student to contribute to a larger conversation dictated in part by the existing document. Second, the final, collectively completed piece was an almost entirely new article. This collaborative effort produced a widely disseminated product, reflected a collective intellectual exercise, and contributed to the wider body of literature and research. Finally, the students observed that the assignment encouraged collaborative engagement and a collective presence through shared resources. It also encouraged a sense of pooled intelligence, proactive learning, and ongoing or dynamic learning.

Blogging/online journaling. Blogging is yet another way to engage students collectively in the learning experience. It is similar to a third space in that it allows the user to share stories and allows the reader to respond, challenge, and engage. Blogs can allow users to engage with one another about the course text or lecture. Students can use blogs to create their own mini-lessons, complete an assignment, or reflect on theological issues. Blogging also allows other students to interact with that particular form of communication and, by extension, the blog author's ideas, thoughts, feelings, and perspectives.

Online journaling is a common outlet for reflection, engagement, and even collaborative learning. One tool for online journaling is Penzu. We typically think of reflection as an individual's journey through a particular period of time or experience, but we used Penzu in this case as a collective reflection device. Participants shared their journal entries with one another, and while it was largely a one-way communication device, Penzu allowed students a "look in," if you will, on their peers' processes of learning and forming.

Other resources. Finally, a tool that Mary has used in some of her courses is TED-Ed, a free educational website for teachers and learners. The TED-Ed website allows students and faculty to create their own "Ted Talk" videos to share and engage with others. In one course, students posted a video in the classroom and then invited fellow students to gather round, virtually, to think together, engage in dialogue with one another, and add additional content to the discussion.

EXPANDING LEARNING ECOLOGIES

These tools and others like them highlight the importance of designing and delivering learning experiences that are collaborative in nature and provide opportunities to expand the scope of one's learning ecology. Stavaredes highlights the importance of collaborative learning as "an opportunity to build interpersonal relationships with peers."[24] From the perspective of those who work in attrition studies, collaborative learning is a key element in retention and satisfaction in any kind of learning venue.[25] Further, many would see collaborative efforts by online members as an important element to building and sustaining a sense of authentic community. Moller et al. noted that "community begins to take shape as the learners communicate, negotiate, and work with one another toward a common goal."[26] We argue that students who collaborate with one another are, in essence, engaging in reciprocal relationship building, and in that engagement with one another experience any number of positive educational and spiritual benefits. We know from both research and anecdotal evidence that in educational venues, collaboration yields enhanced self-esteem, higher course retention rates, higher levels of interest in course work, and increased development of interpersonal skills.[27] The process of collaboration allows for mobilization of "social synergy that resides within a group of co-learners engaged in a dynamic process of shared inquiry. . . . Active engagement and ongoing reciprocity create a community of co-inquirers."[28] The collaborative process of learning embodies a relational approach to community-building found in any number of environments, online or otherwise. The greater the extent to which our students are involved in online activities characterized by collaborative interaction—such as encouraging one another, praying for one another, commenting on one another's discussion board postings, reading and reacting to each other's assignments—the more meaningful their learning is. Those who actively participate in online learning activities rate courses more favorably than those who do not get involved. In fact, one of

---

[24]Tina Stavredes, *Effective Online Teaching: Foundations and Strategies for Student Success* (San Francisco: Jossey-Bass, 2011), 131.

[25]David Arendale, "Review of Successful Practices in Teaching and Learning" (unpublished manuscript, University of Missouri, Kansas City, 1998), 1-24, http://a.web.umkc.edu/arendaled/teachlearn.pdf.

[26]Leslie Moller et al., "Creating an Organic Knowledge-Building Environment Within an Asynchronous Distributed Learning Context," *Quarterly Review of Distance Education* 3 (2002): 47-58.

[27]Arendale, "Review of Successful Practices," 11-13.

[28]Moira Lee, "Experiencing Shared Inquiry Through the Process of Collaborative Learning," *Teaching Theology and Religion* 3, no. 2 (2000): 108-16.

Mary's students admitted that a student holds the final say in "how much you choose to get involved." She honestly reported, "I could have done more."[29] Because of the nature of online learning, there is a greater degree of responsibility on the part of students to participate actively in collaborative and reciprocal exchanges. The issue of intentionality is also at play, not only for faculty and mentors but for students and, in a sense, students' entire network of connections. We encourage our students to make use of those connections in online courses, and we do our best to facilitate that as much as possible by linking reciprocal exchanges in course assignments, group discussions, casual conversations, and deliberative reflections.

We find all of this echoed by Moira Lee, who referenced one participant's observations: "[Collaborative learning] is learning in communities. It is a relationship that exists between the content matter, the community of learners and the teacher."[30] This comment underscores the nature of reciprocity in the learning process. Lee added, "One participant noted that there was a great deal of *one-anothering* that occurred in the classroom experience. That student wrote, 'Our common purpose is to find wholeness in life . . . We learn from each other. There should be one anotherness and togetherness.'"[31] The implication for the community of learners in collaboration is the reciprocal "one-anothering" that exists in shared learning. Collaborative learning as a way of life becomes an intrinsic way of navigating all of one's activities. Collaboration is not just a classroom technique or a boardroom maneuver but rather a way of approaching life and leading others. Lock suggested that for an online community to develop, it must receive the same attention from its members as do face-to-face communities. The formation of community requires a great deal of attention in the area of reciprocity of relationships: "The relationships, the intimacy, the negotiations, and the engagement of participants all influence the evolution of a community. The growth and longevity of a community are directly related to the community meeting the needs of members."[32] At the heart of the issue

---

[29]Mary Lowe, "Assessing the Impact of Online Courses on the Spiritual Formation of Adult Students" (EdD diss., Nova Southeastern University, 2007), 122.

[30]Ibid., 111.

[31]Ibid.

[32]Jennifer V. Lock, "Laying the Groundwork for the Development of Learning Communities Within Online Courses," *Quarterly Review of Distance Education* 3 (2002): 395-408.

is the context within which members of a community live and operate. Kemp addressed this issue in his discussion about contexts and the support system within which a distance learner operates.[33] He included the system of one's church, family, and community as examples of groups that provide social interaction for distance learners.

SUMMARY

One implication of learning ecologies is greater awareness of the intentional social connections we create with one another. From an ecosystems perspective of learning and engagement, we are not as autonomous and independent as we think. In an online classroom, students are not entirely autonomous; they have a support system composed of institutional guidance and contact, family members, mentors, classmates, and professors. We exist in various types of relationships with other people. We are all part of a family, school, church, workplace, voluntary organization, or community. What we do has an effect on others in those social settings. What other people do has an effect on us. There is an inescapable reciprocity between people in a social network. But digital ecologies exist outside classrooms, across the spectrum of our lives and those of students; to this reality we now turn.

---

[33]Steve Kemp, "Learning Communities in Distance Education" (paper presented at the conference of the Association of Christian Continuing Education of Schools and Seminaries, Seal Beach, CA, March, 2002).

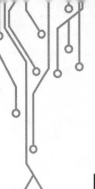

—— SEVEN ——

DIGITAL ECOLOGIES ACROSS
THE DIGITAL LANDSCAPE

Digital natives are more likely to tell stories of how church
emerges whenever Christians gather together for fellowship. In
short, discipleship becomes a way of life that spans the embodied
and the virtual, and this is what it means to be incarnational.

PHILIP MEADOWS, "MISSION AND
DISCIPLESHIP IN A DIGITAL CULTURE"

We like to roast our own coffee beans in a little roaster we keep on our deck. When we first got started roasting green coffee beans from different countries around the world, we were limited in what we could purchase because we did not have a local merchant who would sell us the raw beans. Then the friend who started us on this coffee adventure suggested we try to locate beans online. The Internet opened up a new realm of possibilities in which we could purchase coffee beans from around the world. While we could rarely obtain the Haitian Bleu bean locally, we could find it easily online. Our coffee-roasting experience blends together a real world of roasting times and aromas with virtual-world retailers eager to sell their exotic beans raw and ready to roast. We don't think about our coffee roasting and coffee buying as two distinct realms of existence. We have seamlessly woven together the real and digital realities that provide us a satisfying culinary experience every morning.

It becomes important for us to widen our perspective even further as we embrace our ecological vantage point. While we now appreciate more fully the ecology of learning and the complexity it reveals, it is important to recognize that our learning ecologies occupy a nested place within a larger digital landscape in our culture.

We are just now beginning to appreciate the perceptive insights of thinkers like David Bohm, who help us see that we "need to look on the world as an *undivided whole*."[1] Some of the earliest work conducted in the area of ecological taxonomies was in the early 1900s and related to various scientific disciplines.[2] Scientists determined that an ecology reflects a network in which "the essential properties . . . of a system are properties of the whole that none of the parts alone have."[3] In other words, there are various components to an ecology in which we all live, and yet the individual contributions we make serve a larger purpose. To understand an ecosystem perspective is to essentially understand networks, a dynamic that characterizes much of how we currently live and operate.[4] Raptis et al. observed:

> Ecologies can be described through the network metaphor. . . . The network metaphor implies that every ecology can be understood as nodes that interact among each other through relationships. Furthermore, each node can be perceived as a network itself and the ecological way of thinking focuses on understanding the emergent properties of a network.[5]

There is an inherent and inescapable connection between ecology and network theory. Fritjof Capra underscores this by observing that in nature, or the "web of life," there are nested networks that serve to accomplish specific purposes, none of which are what he considers hierarchical.[6] He understands these to be human projections rather than naturally ordered mechanisms of interconnection. Fritjof's comments occur within a larger conversation about digital ecologies and the need to understand application and assessment for both the user and the practitioner. The 2013 assertion by Ratapis et al. that smart technology is becoming a part of everyday life is

---

[1]David Bohm, *Wholeness and the Implicate Order* (London: Routledge Classics, 2002), 13.
[2]Dimitrios Raptis et al., "What Is a Digital Ecology? Theoretical Foundations and a Unified Definition," *Australian Journal of Intelligent Information Processing Systems* 13, no. 4 (2014), http://people.cs.aau.dk/~jesper/pdf/journals/Kjeldskov-J24.pdf.
[3]Ibid., 1.
[4]Bernard Patten, "Network Ecology," in *Theoretical Studies of Ecosystems: The Network Perspective*, ed. Masahiko Higashi and Thomas P. Burns (New York: Cambridge University Press, 1991), 288-352.
[5]Raptis et al., "What Is a Digital Ecology?," 2.
[6]Fritjof Capra, *The Web of Life* (New York: Anchor, 1996).

even more obvious today. There is a sense in which we confront a manipulated dichotomy between "ubiquitous and mobile computing" and an inclination to differentiate between digital ecologies and everyday life.[7] Raptis et al. define digital ecologies as "*a closed set of digital and non-digital artifacts and a user acting as nodes of a network where its boundaries are specified by an activity and the structure and patterns of organization are either user and/ or designer defined.*"[8]

In a broader perspective, *ecology* describes how the disparate parts of the whole work to accomplish various tasks and responsibilities to achieve an intended outcome. Ecology, properly understood, recognizes not just the parts of the system but *how* those various parts interact with one another to contribute to a larger whole. Some, such as Raptis et al., would argue that digital ecologies are a closed system in which the environment does little to affect outcomes.[9] We, however, agree with others that it is impossible to understand an ecosystem apart from any number of interactions and influences that exist within it.[10] Digital ecologies operate similarly. A digital ecology, in the strictest sense, is a series of interactions between users, available technologies, and the particular culture or context that would define or describe the ecology.[11]

BLURRING THE LINES

John Jewell argues that the search for connection between persons—especially the Net Generation and God—best transpires through the community of God's people as defined in part by the physical, corporate gathering of those people. He notes that "the lines have begun to blur between the real world filled with digital tools and the virtual world created with digital tools."[12] The digital landscape continues to evolve and shift into one

---

[7]Raptis et al., "What Is a Digital Ecology?," 1.
[8]Ibid., 4.
[9]Ibid.
[10]Patten, "Network Ecology," 288-352.
[11]David Meyer, "Creating a More Viable Digital Ecosystem for Faith-Based Humanitarian Groups Working in Kenya to Foster Better Support for Potential Benefactors for Tangible Needs" (master's thesis, Savannah College of Art and Design, 2014), http://ecollections.scad.edu/iii/cpro/DigitalItemViewPage.external?sp=1002562.
[12]John P. Jewell, *Wired for Ministry: How the Internet, Visual Media, and Other New Technologies Can Serve Your Church* (Grand Rapids: Brazos, 2004), 55.

with a new ecology of overlapping contexts. We think this more accurately mirrors life; our social contexts often overlap with one another. In a similar way, digital ecologies are no more rigidly defined than how we live out the rest of our intersecting lives.

Meadows describes this concept in his discussion about digital discipleship and the "remixing of the medium" in such a way as to blur, or even eliminate, divisions.[13] He contends that the idea of incarnation, for what he calls the "digital disciple," is that it can take the form of both virtual and physical realities. He observes that many of these people experience real relationships in either medium and both media work in transformative ways. For Meadows, "These remixed relationships do not necessitate a division between embodied and virtual life: Flesh becomes telepresent through virtual relationships and telepresence is made flesh through face-to-face meetings."[14]

Some would use the term "embodied" to differentiate the physical from the virtual self, but the idea that embodiment runs contrary to the virtual self is a dichotomization we are not alone in rejecting. People who come to the virtual portal do not, contrary to critics of virtual engagement, somehow leave their bodies, minds, and emotions at the front door of virtual experiences. The one who provides comfort and nurture to those physically present is the same person who seeks answers or provides comfort to others online. Hess argued that online education does not, in fact, lead to disembodied learning.[15] The individual is very much interacting with the medium of technology through the use of the mental, physical, emotional, and spiritual dimensions. Moreover, she noted that there is nothing inherently engaging about physical space. In fact, she countered that some of what instructors teach in the traditional classroom often appears to students as disconnected, or disengaged, from the world in which students live and minister.

This mirrors what Meadows observed about digital immigrants. He suggests they are more likely to view structures such as the church sanctuary as a place strictly designed for Christian fellowship and worship. Those that he

---

[13]Philip Meadows, "Mission and Discipleship in a Digital Culture," *Mission Studies* 29 (2012): 163-82.

[14]Ibid., 175.

[15]Mary E. Hess, "Attending to Embodiedness in Online, Theologically Focused Learning" (unpublished paper, Luther Seminary, 2000), www.luthersem.edu/mhess/dayton.pdf.

refers to as "digital natives," on the other hand, are more easily able to fluc-
tuate between coffee houses, church house groups, and online sites as ways
to engage each other in spiritual fellowship. Natives view these links as

> connective spaces which can flow together and where real people can make
> and sustain transforming relationships with God and one another. More im-
> portantly, church can exist between these spaces as spiritual friendships that
> become deeply embedded in the ordinary routines of daily life.[16]

Connective spaces, in this context, are those digital ecologies in which there
is fluctuating interaction between various systems within that space. Those
spaces serve to create what many believe to be deeply transformative and
highly relational experiences and encounters.

EMBODIED VIRTUALLY

Much of the literature about virtual communities from a theological perspective has centered largely on the practice of distinguishing embodied
from virtual as a way of elevating the embodied community over the digital
community. An ecological perspective helps us understand that the divide
is not as sharp as critics of online learning assert. As we see in nature, plants
depend on a variety of elements from a number of systems for their survival.
We would be remiss to think that plants don't occupy a certain space in their
need of light from the solar system or nutrients from the soil. There are
nutrient exchanges between ecological habitats that enable each to grow and
thrive. Similarly, humans exchange resources across various sectors of our
human ecologies. Urie Bronfenbrenner laid this out succinctly in his descriptions of the developing person in relation to her or his social ecologies.
In his perspective, everything we do stems from the notion of being "inextricably embedded . . . in particular settings" with the understanding that
the boundaries of each of those settings are porous and fully embedded
within the other.[17] He then notes that "there is always an interplay between
the psychological characteristics of the person and of a specific environment;
the one cannot be defined without reference to the other."[18] Bronfenbrenner

---

[16]Meadows, "Mission and Discipleship," 175.
[17]Urie Bronfenbrenner, *Making Human Beings Human* (Thousand Oaks, CA: Sage, 2005), 146.
[18]Ibid.

uses the taxonomy of systems to define his ecology of human development with terms like "microsystem," which represents the direct environment within which interpersonal relationships form. He uses the family as a prime example of this particular system, which highlights the role of reciprocity in development. The "mesosystem" is what people experience in their exchanges across multiple contexts. He describes this as the "linkages and processes taking place between two or more settings containing the developing person" as a way of identifying systems within systems.[19] The relationship between the home and school contexts is a good example of this type of system. The third level is the "exosystem" spheres which may never connect directly, but which combine to influence an individual. Bronfenbrenner uses the example of the effect on a child between the home environment and a parent's work or school. The "macrosystem" reflects the culture in which the developing person grows and develops. This may represent socioeconomic factors, ethnic variables, or religious norms. Finally, the "chronosystem" suggests that we undergo formation, in part, by transitions or life changes that occur in varying life stages. This ecological view helps us to understand the effects of various systems and contexts that often overlap and interact throughout a person's life.

Diana Garland's work in family ecology echoes some of the same findings that Bronfenbrenner shares. She observes,

> Just as we cannot understand individuals without attending also to the family relationships that have nurtured and shaped them, neither can we understand families without looking beyond them to the social environment that nurtures, shapes, and is influenced by the family. The ecosystemic perspective uses ecology as a metaphor for the relationships human systems (families) have with their physical and social environments.[20]

Garland adds to these observations terms such as *habitat, adaptation,* and *stress* to describe the various family dynamics that serve to influence and be influenced by the systems in place. A family's habitat, for example, provides a physical space within which its members can thrive or regress, depending on variables within that system. Diane Leclerc and Mark Maddix

---

[19]Ibid., 148.

[20]Diana R. Garland, *Family Ministry: A Comprehensive Guide* (Downers Grove, IL: IVP Academic, 2012), 242.

echo Garland's perspective on the relational component of spiritual maturation within families. Leclerc and Maddix note that "the relational qualities
of the family can mirror the relational power of the triune God for redemptive purposes."[21] The simple act of table fellowship, whether in a
kinship structure or one that reflects a family of faith, underscores the significant impact of reciprocal spirituality. According to Leclerc and Maddix,
the simple act of eating together, inviting God's blessing on the shared meal,
and sharing of conversation with one another are ways to nurture faith. An
ecological perspective on virtual presence helps us recognize the embedded
reality of human existence in human ecologies large and small.

Jewell's claim that "virtual community does not have 'real presence' . . .
where real people gather as the body of Christ" truncates the larger perspective of how ecosystems serve to form our understanding of growth and
development.[22] When viewed strictly from that perspective, he may be on
point, but we believe this is a myopic viewpoint; it only focuses on one element of the digital ecology.

PAUL'S MEDIATED PRESENCE

Quoting the Greek philosopher Proclus (412–485 AD), Doty indicates that
the widespread belief in the first century was that a letter should be written
"to someone not present *as if he were present.*"[23] This is what Keck concluded
from his analysis when he wrote, "Paul regarded his letters as stand-ins for
his own presence. When a letter from Paul was read, it was as though *Paul
himself were speaking.*"[24]

Mitchell takes this evidence further when she asks, "Is it not more
likely the case that in certain instances Paul sent envoys or letters (or
both) to represent him because he thought that they might be *more effective than a personal visit* in dealing with a particular situation that was
facing a church?"[25] Richard Ward offers this insight: "When rendered

---

[21]Diane Leclerc and Mark A. Maddix, *Spiritual Formation: A Wesleyan Paradigm* (Kansas City, MO:
Beacon Hill Press, 2011), 185.

[22]Jewell, *Wired for Ministry*, 61.

[23]William G. Doty, *Letters in Primitive Christianity* (Philadelphia: Fortress, 1973), 12. Emphasis mine.

[24]Leander Keck, *Paul and His Letters* (Philadelphia: Fortress, 1979), 18. Emphasis mine.

[25]Margaret M. Mitchell, "New Testament Envoys in the Context of Greco-Roman Diplomatic and
Epistolary Conventions: The Example of Timothy and Titus," *Journal of Biblical Literature* 111, no.
4 (1992): 641-62.

orally, the form of the letter is transformed into a presence that is embodied by the reciter."[26] However, Leea Johnson argues that although envoys are important, as Mitchell ably demonstrates, it was the content of the letter, not the person delivering it, that most directly embodied Paul's mediated presence.[27]

Within the serial exchange of letters between Paul and the church at Corinth, we have a biblical case study of the role Paul's letters played in the churches he founded. Although we have two extant letters from this exchange, both written by Paul, we know from 1 and 2 Corinthians that Paul wrote other letters to the Corinthian church and that the Corinthians wrote letters to him. First, we must note that Paul never expresses a desire or eagerness to see the Corinthians, in contrast to what he says in most epistles to other churches. The only place in his correspondence that may be construed in this way, 2 Corinthians 8:16-17, actually relates to the earnest desire of Titus to visit Corinth, not Paul. In many of Paul's other letters, he explains to the recipients why he was hindered from visiting the church, but he offers no such explanation in his Corinthian correspondence. Instead, Paul explains his absence from Corinth based on personal preference. Why would Paul prefer not to be present in Corinth, especially if personal presence is the most effective form of personal communication? The answer may be that Paul is intentionally distancing himself from the church to prompt the church into a reconciled relationship with him. Johnson argues that Paul deliberately avoided a personal visit and instead preferred to mediate his presence through his letters because his presence in a previous visit created the breach in their present relationship.[28] In fact, Paul writes to the church in Corinth to tell them that instead of making a painful visit, "I determined this for my own sake, that I would not come to you in sorrow again. . . . For out of much affliction and anguish of heart I wrote to you with many tears; not so that you would be made sorrowful, but that you might know the love which I have especially for you" (2 Cor 2:1-4). Paul reveals his heart to them when he writes in 2 Corinthians 12:20-21, "For I am afraid that perhaps when

---

[26]Richard F. Ward, "Pauline Voice and Presence as Strategic Communication," *Semeia* 64 (1994): 95-107.

[27]Leea Johnson, "Paul's Epistolary Presence in Corinth: A New Look at Robert W. Funk's Apostolic 'Parousia,'" *The Catholic Biblical Quarterly* 68, no. 3 (2006): 481-501.

[28]Ibid.

I come I may find you to be not what I wish and may be found by you to be not what you wish, . . . I am afraid that when I come again my God may humiliate me before you."

SPIRITUAL FORMATION AND DIVINE PEDAGOGY

Paul claims to have contributed to the spiritual well-being of his recipients in many of his letters. Through his epistles, Paul was able to carry out ministry and influence the spiritual development of individuals in his churches. The reason for this spiritual impact with mediated communication was the empowering ministry of the Holy Spirit working through the written words of Paul's epistles.

In addition to Paul's use of letters as a way to make himself present in spirit while absent in body, we can reference "divine pedagogy," a concept that flourished during the time of the early church fathers, as another way to counter the arguments against online learning in Christian education. This concept originated among the early church fathers, as Gresham says, "To describe the historical process by which God gradually and progressively taught and prepared the human race for the fullness of revelation in the coming of Christ."[29] Gresham also points out that God as divine pedagogue mediates his presence and revelation to his creatures. God mediated his presence indirectly through the Spirit, through the Ark and Tabernacle, through the glory cloud, through the Temple, and, most dramatically, through the incarnation of his beloved Son. Moreover, God mediated his revelation through signs and wonders, human instruments such as prophets and apostles, the written word, and the Living Word, Jesus Christ.

Second, divine pedagogy recognizes that God adapts his revelation to his recipients through human language (e.g., Greek and Hebrew) and human literary forms, such as poetry, narrative, history, parable, and proverb. Finally, divine pedagogy understands that God accommodates his revelation to human capacity by its progressive and developmental character. Any wise teacher knows that students cannot absorb all they need to know in one massive downloading of material. We spread out what we want to teach over

---

[29]John Gresham, "The Divine Pedagogy as a Model for Online Education," *Teaching Theology and Religion* 9, no. 1 (2006): 24-28.

a protracted period of time in accordance with the developmental trajectory of our students. The concept of divine pedagogy reminds us that God used mediated forms of communication and empowered their effectiveness through the disembodied ministry of the Holy Spirit.

If Paul practiced a form of mediated instruction through the use of letters to his churches but was able to make himself present to his churches through those letters, then his use of a mediated technology was not disembodied or impersonal. Paul incarnated his very presence to his churches through the letters he wrote that were read aloud to his congregations by his emissaries. Further, we have seen that it is possible to construe from Paul's letters to the Corinthians that he considered the epistolary medium a form of communication superior to his personal presence in the community. In addition, we have seen that through this form of mediated presence, he and other New Testament apostles were able to promote the spiritual formation of their recipients. Finally, we have seen that the early church considered God's pedagogical method to involve some form of mediated communication. God mediated his presence primarily through the Holy Spirit and mediated his revelation through the written word—including the use of letters that embodied apostolic presence—so that God's people experienced ongoing transformation into the fullness of Christ.

DIGITAL LANDSCAPES

The digital revolution affects nearly every segment of our culture. Social media have reshaped every digital landscape we once knew. The students coming to our classrooms interact with other people and get their information in a significantly different way than did previous generations. According to one report, "Over the past sixty days, more information has been uploaded to YouTube than if ABC, NBC, and CBS had been broadcasting around the clock since 1948."[30] YouTube has revolutionized how millennials are getting their information and sharing information about themselves. Millennials, a generation defined by the Pew Research Center as "those aged 18-34 in 2015, now number 75.4 million," are the most dominant in terms of

---

[30]Phil Schubert, "Grasping the Realities of Educating in the Digital Age," *EDUCAUSE Review,* last modified April 7, 2011, www.educause.edu/ero/article/grasping-realities-educating-digital-age.

national demographics.[31] We use the terms "millennial" and "digital native" interchangeably to refer to those who represent a certain group of people identifying with a culture having its own language, practices, and understanding of relationships, spirituality, and global awareness. Marc Prensky applies the term "digital native" to his description "'native speakers' of the digital language of computers, video games and the Internet."[32] He differentiates the digital native from "digital immigrants," or those of us who "at some later point in our lives, became fascinated by and adopted many or most aspects of the new technology."[33] This terminology reflects a larger narrative about the divide in our culture between digital generations. It also reflects a kind of ecology in which digital practices and disciplines have largely merged with what some might consider nondigital applications. Many millennials see the two in a much more seamless manner than we may have previously understood.

It is now apparent that digital forms of interaction have shifted from largely informational to increasingly relational ways of interacting without the loss of the information. In fact, according to one report, nearly 90% of people ages of 18 through 24 check social media before getting out of bed in the morning.[34] Those reporting on this study maintain that "the most common sentiment regarding smartphone is one of 'connectedness,' far surpassing 'overwhelmed,' 'stressed out,' 'burdened/anxious,' or 'lonely.'"[35] While many use social media to check news and other things, the overwhelming majority are still using it to maintain connections with one another. In her study of the social lives of teenagers, Danah Boyd made this startling discovery: "Most teens aren't addicted to social media; if anything, they're addicted to each other."[36] That is, teenagers use social media to connect socially in an environment "free from adult surveillance."[37] In her interviews with

---

[31]Richard Fry, "Millennials Overtake Baby Boomers as America's Largest Generation," Pew Research Center, April 25, 2016, www.pewresearch.org/fact-tank/2016/04/25/millennials-overtake -baby-boomers/.

[32]Marc Prensky, "Digital Natives, Digital Immigrants," *On the Horizon* 9, no. 5 (2001): 1-6.

[33]Ibid., 2.

[34]Allison Stadd, "79% of People 18-44 Have Their Smartphones with Them 22 Hours a Day," *Adweek,* April 2, 2013, www.adweek.com/socialtimes/smartphones/480485.

[35]Ibid.

[36]Danah Boyd, *It's Complicated: The Social Lives of Networked Teens* (New Haven, CT: Yale University Press, 2014), 80.

[37]Ibid., 86.

these highly interconnected teens, she discovered that they would much rather gather together with their friends in person. She recounts her experience attending Friday night high school football games where all the students sit together in the stands talking to each other face-to-face and only used their phones to take pictures of the team, cheerleaders, or homecoming royalty.

Digital ecologies. The presence of digital ecologies is already shaping how people develop spiritually. According to one body of research, "70% of practicing Christian millennials read Scripture on a screen."[38] Millennials are going online to find answers to their spiritual questions. More than 50% of Christian millennials watch online videos with faith content. An increasing number of millennials are searching online for a place of worship, and some are attending that place online. House of Prayer by Second Life is one of the fastest-growing sites for those who don't find real-life participation part of their desired experience. One of our colleagues, Christine Vintinner, made a focused attempt at understanding how one inhabits and participates in Second Life, a Bible-believing group of Christians gathered virtually rather than bodily. She found that participants can experience "lasting friendships, rewarding activities, and diversity," which is what keeps them returning to the site.[39] Further, what they experience in their normal, daily lives accompanies them online in Second Life; rather than a game, as some might suggest, participation in Second Life is similar to what they would encounter in their offline lives. Vintinner adds that the House of Prayer Church is similar to chatrooms but includes a visual element, and nearly every denomination is represented in Second Life. Members gather to hear sermons, worship, and engage in Bible study groups that meet every week. They participate in prayer groups and support teams to help with the challenges of everyday life. Finally, they observe memorials, remembrances, and holiday services, much like their offline counterparts. For these and others who find the online platform to be a sacred space and one in which they find belonging and opportunities to contribute, these are spiritually formative experiences.

---

[38]Barna Research, "How Technology Is Changing Millennial Faith," last modified October 15, 2013, www.barna.org/barna-update/millennials/640-how-technology-is-changing-millennial-faith# .Ul9dGFN21qM.

[39]Christine Vintinner, email message to authors, August 24, 2013.

Barna researchers found that "14% of Millennials say they search to verify something a faith leader has said. A striking 38% of practicing Christian Millennials say the same."[40] They add that nearly 30% of all millennials are going online to find spiritual content. Digital natives are engaging in real-life worship with their smartphones. For them, faith is experiential and interactive, and in order to do this most effectively, they engage in matters of faith and worship with their digital devices. Users of Twitter and Instagram do not just use them for tweeting updates or posting pictures; rather, a growing number of digital natives use various social technologies to talk about matters of faith. Barna reports that "40% of Christian Millennials say they participate in online conversations about faith, and the same number say they blog or post comments on blogs about spiritual matters."[41]

In 2001, the Barna Group predicted that by 2010, close to fifty million people would depend on the Internet for their spiritual interaction, adding that more than two-thirds of those polled expected to use the Internet regularly "to seek or engage in specific types of religious experiences" as time passed.[42] Those activities included listening to religious teaching, purchasing religious products, and reading devotionals online. According to a report published in 2014 by the Pew Research Center for Religion & Public Life, nearly sixty million, or one-fifth of all Americans, "share their religious faith online, about the same percentage that tune in to religious talk radio, watch religious TV programs or listen to Christian rock music."[43] Moreover, the Barna Group provides further evidence that pastors are more open and willing to embrace the contribution of online experiences to "faith assistance or religious experiences."[44] According to their findings, nearly 90% of pastors today say "they think people in their area would find it acceptable for their church to provide faith assistance or religious experiences to people

---

[40]Barna, "How Technology Is Changing Millennial Faith."

[41]Ibid.

[42]Barna Group, "More Americans Are Seeking Net-Based Faith Experiences," last modified May 21, 2001, www.barna.com/research/more-americans-are-seeking-net-based-faith-experiences.

[43]Pew Research Center, "Religion and Electronic Media: One-in-Five Americans Share Their Faith Online," last modified November 6, 2014, www.pewforum.org/2014/11/06/religion-and-elec tronic-media/.

[44]Barna Group, "Cyber Church: Pastors and the Internet," last modified February 11, 2015, www .barna.com/research/cyber-church-pastors-and-the-internet/.

through the Internet."[45] Nearly the same number of pastors take the position that "it is theologically acceptable for a church to provide faith assistance or religious experiences to people through the Internet."[46]

The population groups identified by John Horrigan and others are similar to some descriptions about faith activity online. Elena Larsen reported that "28 million Americans have used the Internet to get religious and spiritual information and connect with others on their faith journeys."[47] She maintained that the figure represents an increase from 21% to 25% of Internet users seeking online faith-based experiences. Moreover, Larsen noted, "Between 19 million and 20 million people have gone online to get religious or spiritual material."[48] Her analysis found that "more than 3 million people a day get religious or spiritual material [online], up from 2 million that we reported last year."[49] Larsen's study revealed that users go online more often for faith-based experiences than for other services, such as online banking, stock trading, or web auctions. Larsen noted that "41% of Internet users, many of whom had never considered themselves online spiritual seekers, said they sent or received email prayer requests."[50]

Digital hierarchies. One of the reasons social media sites have gained popularity among digital natives is that they find this medium to be much more egalitarian than the structures they find in place with real life. Social media often get credited with toppling governments, empowering those who may be marginalized, contributing to the success of political campaigns, and enhancing grassroots activism, all of which act as an extension of real life activities. Felicia Wu Song points out that online communities "have proven to play extremely pivotal roles in aiding activists in social movements."[51] Social or political activism is not the only thriving activity, however. Song goes on to note that "online communities have also uniquely

---

[45]Ibid.

[46]Ibid.

[47]Elena Larsen, "CyberFaith: How Americans Pursue Religion Online," Pew Research Center, last modified December 23, 2001, www.pewinternet.org/2001/12/23/cyberfaith-how-americans -pursue-religion-online/.

[48]Ibid.

[49]Ibid.

[50]Ibid.

[51]Felicia Wu Song, *Virtual Communities: Bowling Alone, Online Together* (New York: Peter Lang, 2009), 4.

functioned to facilitate the public's immediate response to major events in the world."[52] Christina Neumayer and Celina Raffl echo this by noting that "cyber-space is not a sphere of its own, distinct from real life, but an expression of societal structures that are to some extent transferred to the virtual space, and vice versa."[53] They argue that there is a seamlessness or reciprocity in the activism that surfaces in real life as well as cyber spaces. Song underscores this interplay between culture and technology by noting that "because the Internet is constituted by a cultural and normative environment, the structural and discursive features of online communities can illuminate significant cultural conceptions and patterns that bring into focus what a society privileges, dismisses, or takes for granted."[54]

SUMMARY

The larger picture in this description of digital ecologies is that they fully embed our lives. While some may prefer to come at the issue of digital ecologies in a compartmentalized view, there seems to be, based on the evidence provided in this chapter, a greater sense of overlap. This issue of overlap drives our perspective of and practice of spiritually formative opportunities. We see spiritual formation as something we integrate into who we are and what we do rather than segmenting it from the rest of our lives. Technological prognosticators, such as Michio Kaku, anticipate that "the destiny of computers—like other mass technologies like electricity, paper, and running water—is to become invisible, that is, to disappear into the fabric of our lives, to be everywhere and nowhere, silently and seamlessly carrying out our wishes."[55] There is a sense in which a "both/and" principle applies; our culture has become enmeshed in digital technology to the degree that we no longer define barriers or spatial limitations. To some degree, this is a helpful way of understanding that our participation in

---

[52]Ibid., 5.

[53]Christina Neumayer and Celina Raffl, "Facebook for Global Protest: The Potential and Limits of Social Software for Grassroots Activism" (paper presented at the Community Informatics Conference, ICTs for Social Inclusion, July 2008), www.researchgate.net/publication/240642461_Facebook_for_Global_Protest_The_Potential_and_Limits_of_Social_Software_for_Grassroots_Activism.

[54]Wu Song, *Virtual Communities*, 8.

[55]Michio Kaku, *Physics of the Future: How Science Will Shape Human Destiny and Our Daily Lives by the Year 2100* (New York: Random House, 2011), 23.

offline social contexts and networks have the potential to form us, and we bring those connections to bear in online connections, and vice versa. At the same time, we have a responsibility as people made in the image of Christ to honor and facilitate the formation of the body of Christ in a way that seeks to enact the fullness of Christ with one another. Part of that responsibility encourages us to honor one another in each of the ecologies in which we find ourselves. Our participation in reciprocal exchanges with others, empowered by the Holy Spirit, instigates continued growth and transformation. In the next chapter, we will explore social ecologies' power to influence us and how that power can help us influence others.

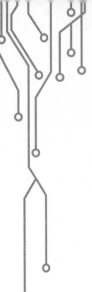

—— EIGHT ——

SOCIAL NETWORKS
AND THE POWER OF
RECIPROCAL INFLUENCE

*If nature reflects the image of God and the human religious quest
is embedded within and shaped by the physical and biological
systems of our planet, then a complete understanding
of spiritual formation requires a consideration of the
ecological dimensions of Christian spirituality.*

Timothy H. Robinson, "The Role of
Nature in Spiritual Formation"

In January 2010 we were leading a medical mission team to Haiti when the
earthquake struck and devastated the country. Our medical team switched
from preventative medicine to emergency medicine as they treated sur-
vivors who had sustained lacerations, broken bones, and internal injuries.
While they were attending to the victims of the earthquake, Mary and I were
making contacts with Christian relief organizations to get medicines and
other medical supplies into the country. We were also trying to make contact
with family members of our medical team and our local church, who spon-
sored us. The only way we could keep in touch was through Facebook. For
some reason, the Internet connection at the mission house where we stayed
remained intact in spite of the destruction around us. Through Facebook,
we were able to give regular updates to our own families and our church
family so that they knew we were safe and trying to get back home to them.

At the time, we had a relative of a friend working in the White House who asked if any military transport planes would be landing at the airport near us. We discovered that the runway was too short and thus we had to eliminate this option for getting our team back to their families. It was through Facebook that we made contact with our pastor, who had received a phone call from another pastor, who relayed information about a mission agency flying in medical supplies to the airport near our location. We were able to reach them, add our medical needs to their list, and make arrangements to have our team flown to the Bahamas on their empty planes. The social ecology connections of everyone involved were firing in all directions to help alleviate suffering and retrieve our team and reunite us with anxious family members and our church family.

When we see the term "ecology," we usually associate it with science and nature. Using a framework of biology helps us understand that an ecosystem approach represents a complex system of interactive, dynamic, and bidirectional influences that typically foster growth and development. Roderick McKenzie, one of the pioneers in understanding human ecology, points to the "fundamental interrelatedness of all living things" in nature as a way of describing and understanding human ecology.[1] He understood human ecology as the study of "the relations of [hu]man to [hu]man."[2] Alfred North Whitehead noted that "the various parts of nature are so closely interdependent, so densely woven into a single web of being, that none may be abstracted without altering its own identity and that of the whole."[3] For Whitehead and a number of twentieth-century scientists, individualism was largely misplaced in the economy of nature and how things grow and develop. Many of these scientists were responding to what they viewed as "the ethic of unrestrained individualism" promoted in the social sciences and other disciplines.[4] Some believed that rugged individualism had contributed to a fragmentation of culture and community, while nature, by contrast, represented a unified "closely knit community."[5] For these and others,

---

[1] Roderick D. McKenzie, *On Human Ecology* (Chicago: University of Chicago Press, 1968), xii.
[2] Ibid., 40.
[3] Donald Worster, *Nature's Economy: A History of Ecological Ideas*, 2nd ed. (New York: Cambridge University Press, 1994), 317.
[4] Ibid., 319.
[5] Ibid., 325.

there was incompatibility between unbridled independent individualism and "organismic interrelatedness" observed in nature.[6]

DEFINING A SOCIAL NETWORK

The study of social networks in the United States began with the publication of Jacob Moreno's *Who Shall Survive? A New Approach to the Problem of Human Interrelationships*.[7] His work gave rise to the field of sociometry, which morphed into the study of social networks. Urie Bronfenbrenner built his concept of the ecology of human development embedded in social networks and instigated by reciprocal interactions upon Moreno's work.[8] Bronfenbrenner opened his article with this context-setting statement:

> The emphasis of modern educational theory on the socio-emotional aspects of human growth has imposed the necessity of devising techniques for evaluating the degree and character of social development. The problem has been complicated by the fact that social development applies not only to the individual but also to the social organization of which he [or she] is a part.[9]

The experimental work of Jeffrey Travers and Stanley Milgram to test what they termed "the small world problem," but which in popular culture is known as "six degrees of separation," increased our awareness of humanity's social connectedness.[10]

Christakis and Fowler provide the simplest definition of a social network when they suggest that "a social network is an organized set of people that consists of two kinds of elements: human beings and the connections between them."[11] While they identify connections as the starting

---

[6]Ibid., 321.

[7]Jacob Moreno, *Who Shall Survive? A New Approach to the Problem of Human Interrelationships* (Washington, DC: Nervous and Mental Diseases Publishing Company, 1934).

[8]Urie Bronfenbrenner, "A Constant Frame of Reference for Sociometric Research," *Sociometry* 6, no. 4 (1943): 363-97.

[9]Ibid., 363.

[10]Albert-László Barabási, *Linked: How Everything Is Connected to Everything Else and What It Means for Business, Science, and Everyday Life* (London: Plume, 2003), 25-40; Jeffrey Travers and Stanley Milgram, "An Experimental Study of the Small World Problem," *Sociometry* 32, no. 4 (1969): 425-43.

[11]Nicholas A. Christakis and James H. Fowler, *Connected: The Surprising Power of Social Networks and How They Shape Our Lives* (New York: Little, Brown, 2009), 13.

point for the formation of a social network, they explain that "what actually flows across the connections is also crucial."[12] The term they use to describe what transpires between connected people is "interactions."[13] Social interactions describe "personal relationships . . . that are embedded (strongly connected) in stable networks of trust and reciprocity."[14] The combination of connections between people and the interactions that take place through those connections gives rise to the potential of social contagion or interpersonal influence.[15]

The issue of social networks came to light most significantly after one of the worst crises in American history: the terrorist attacks perpetrated on September 11, 2001. The use of social network analysis became a more prominent tool in the months following the act of terrorism as a way of not only tracking but understanding the motivation toward violence of network members.[16] Those who study social framework or network analysis see far less randomness between terrorist activities and network placement and participation. In fact, Perliger and Pedazhur advocate that "the formation of the social framework responsible for the violence [can] precede the cause of the violence."[17] In other words, understanding the social connections within a network may help us understand the reason for the perpetration. The terrorists who carried out the attacks on New York and other sites were not random perpetrators. They were connected to one another, to the leader, and to the Al Quaida network through a series of relationships.

Persons often expand their social networks through various forms of social media, fostered by interactions, emotional support, frequency of contact, and other developmental features. The advent of the Internet changed the landscape of how we understood the concept of community devoid of physical proximity. Campbell and Garner maintain that advances in technology gave way to a new understanding of social networks and likely

---

[12]Ibid., 21.
[13]Ibid., 26, 32.
[14]Ibid., 161.
[15]Ibid., 16, 22, 105-12.
[16]Arie Perliger and Ami Pedahzur, "Social Network Analysis in the Study of Terrorism and Political Violence," *Political Science and Politics* 44, no. 1 (2011): 45-50.
[17]Ibid., 46.

"birthed social network analysis."[18] They add that what emerged from this new perspective was the idea that

> communities are in their essence social structures and not spatial or geographic structures such as neighborhoods. Social network analysis is a method used to identify a set of nodes (which can be persons, groups, or organizations) and the ties between all or some of them in order to understand the social structures that emerge from the network of relationships.[19]

This new paradigm allowed social scientists to understand how relationships formed online can in fact reciprocate, grow, and develop in a way that is equivalent to offline relationships. Campbell and Garner underscore this point by noting that "the network metaphor has become an important frame for describing not only the function of online communities but also the nature of community offline."[20] A "social network," then, refers to the connections people have with each other, the interactions that transpire across the connections, and the influence that results from these connective interactions.

SOCIAL CONNECTIONS

The first identified component of social network function explores the fact that some kind of social connection must exist between two or more people for a network structure to emerge. What emerges from our social connections with other people is a network topology that reflects a pattern of ties between individuals.[21] Social connections take many different forms. We have family, friendship, work, church, and educational connections. The person with the most social connections in a social network occupies a place of centrality and is the "hub" of the social network.[22] We commonly refer to a person with many connections as a "social butterfly." These individuals possess the ability to befriend a large number of people and stay in contact with them using a variety of means. They also occupy a place of centrality in a social network where they "have quick access to information circulating

---

[18]Heidi A. Campbell and Stephen Garner, *Networked Theology: Negotiating Faith in Digital Culture* (Grand Rapids: Baker Academic, 2016), 7.
[19]Ibid., 7.
[20]Ibid.
[21]Ibid., 16.
[22]Barabási, *Linked*, 55-64.

in the network" and they "may control the circulation of information" in the network.[23] But the "hub" person in a social network is not the only contagious person. Vladimir Barash analyzed the dynamics of social contagion and discovered that "any individual in a population can be infected by any other individual at any time."[24] Further, this means that "infected acquaintances exert as much influence as close friends."[25] In fact, Mark Granovetter discovered that our weak tie acquaintances who operate on the fringe of our social networks display a "strength of weak ties" phenomenon others had overlooked.[26] He found that among blue-collar and white-collar workers, the majority were hired in their current position through information gleaned from weak tie acquaintances who were only "marginally included in the current network of contacts."[27]

Our social connections produce a social network that reflects either uniformity or diversity. Social network analysts use the term *homophily* to describe social networks where we connect with others like us. The word *homophily* literally means "the love of being alike" but most often comes to expression in the adage "birds of a feather flock together." Most people tend to associate with people who are like them in some respect (socioeconomic level, educational attainment, political persuasion, etc.). Some people like more diversity among their friends and acquaintances and thus deliberately form connections with those who are different in many of the areas identified above. However, as Alain Degenne and Michel Forsé report, "Despite some slippage over the past 30 years, homophily remains significantly more probable than heterophily."[28] This is probably because "people tend to select friends from among those in their own social environment where most introductions to other people occur."[29] Ecological diversity in natural ecosystems is a given. Ernest

---

[23]Wouter de Nooy, Andrej Mrvar, and Vladimir Batagelj, *Exploratory Social Network Analysis with Pajek* (Cambridge: Cambridge University Press, 2005), 133.

[24]Vladimir Barash, "The Dynamics of Social Contagion" (PhD diss., Graduate School of Cornell University, 2011), 26.

[25]Ibid., 80.

[26]Mark S. Granovetter, "The Strength of Weak Ties," *American Journal of Sociology* 78, no. 6 (1973): 1360-80.

[27]Ibid., 1371.

[28]Alain Degenne and Michel Forsé, *Introducing Social Networks* (London: Sage, 2006), 34.

[29]Ibid.

Callenbach writes that "in general, ecosystems with more DIVERSITY are thought better able to survive for a long time because they contain more species in complex interactions."[30] What social scientists have yet to demonstrate (though we believe they will) is that the same is true for social ecosystems. The more diverse the natural ecology, the greater its ability to withstand negative disturbances, infections, diseases, and other natural calamities, like lightning strikes that start forest fires. Social systems that reflect diversity in our social connections with others will be similarly robust. Spiritual ecologies are no different. A missional reading of Scripture enables us to see that God's intent from the beginning was to enlarge the covenant community beyond the original family of Abraham to all the families of the earth (Gen 12:3). The ultimate outcome of the witness of Israel and the church is provided by John's vision of "a great multitude which no one could count, from every nation and *all* tribes and peoples and tongues" (Rev 7:9).

SOCIAL INTERACTIONS

Social interconnections make possible social interactions between persons linked through a variety of social ties. Social networks function like natural ecologies, especially in regard to the "interrelated and interdependent" nature of individuals to each other and the human communities they inhabit.[31] As in various components of an ecosystem, the reciprocity that a person encounters through bidirectional social interaction and influence is critical for the continued growth and development of that person. The result of that interactive environment can be transformative in every dimension of our lives because it follows the created order of how things grow and develop. A plant that is part of the natural ecology doesn't simply grow one dimensionally but multidimensionally. The environment can affect the stem, the leaves, and the rest of the plant, negatively or positively. In response, the plant provides, in a reciprocal sense, necessary components to its environment. It can provide nutrition to the soil, oxygen to the air, and food for other animals. There is reciprocal interaction in a social network ecology just as there is in a natural ecology. When we engage with others in our

---

[30]Ernest Callenbach, *Ecology: A Pocket Guide* (Berkeley: University of California Press, 2008), 41.
[31]McKenzie, *On Human Ecology*, 40.

social ecosystem, there can be both positive and negative effects of that interaction. When we forgive one another, we not only alleviate unnecessary burdens of our own but we allow the other person to re-enter a right relationship with us and others. When we encourage others, they can reciprocate by engaging in kind deeds or words. If growth and development in a human ecosystem mirrors growth and development in the natural order, it stands to reason that spiritual formation follows similar patterns of development. A human ecology consists of interconnected individuals interacting with one another, thereby influencing each other's human growth and development.

SOCIAL INFLUENCE

Social network analysis suggests that people connected in a social network influence one another in a variety of ways. Charles Kadushin differentiates "social networks" from "groups" in that the former is a "set of relationships."[32] Christakis and Fowler distinguish social networks from simply a group of people.[33] They argue that the connections between people in a social network have more of an influence on behavior or affect than the individuals themselves. For Christakis and Fowler, "The ties explain why the whole is greater than the sum of its parts."[34] The value of reciprocity between connected and interactive social ties is that our social networks may help us accomplish what we cannot do on our own. Connectedness, Christakis and Fowler note, "carries with it radical implications for the way we understand the human condition."[35] We cannot, in good faith, promote the value of isolation and marginalization as individuals when we understand the value of social networks.

One of the things that Christakis and Fowler point out about a social network is its ability to act as an agent of contagion.[36] Like a disease, various beliefs, behaviors, attitudes, and norms can spread through social networks.

---

[32]Charles Kadushin, *Understanding Social Networks: Theories, Concepts, and Findings* (New York: Oxford University Press, 2012), 14.

[33]Nicholas A. Christakis and James H. Fowler, *Connected: The Surprising Power of Our Social Networks and How They Shape Our Lives* (New York: Little, Brown, 2009).

[34]Ibid., 9.

[35]Ibid., 31.

[36]Ibid., 105-12.

Evidence from their work and others indicates that alcoholism, happiness, depression, suicides, and even obesity spread from one person to another through social connections.[37] Adam Kramer et al. conducted a large-scale study of network contagion involving the spread of emotions (positive and negative) through a social network.[38] The participants in the study numbered 689,000 online users of Facebook. The experimental study found that "emotions expressed by friends, via online social networks, influence our own moods . . . providing support . . . that emotions spread via contagion through a network."[39] One fascinating piece of evidence emerged from this study when they found that "although nonverbal behavior is well-established as one medium for contagion, these data suggest that contagion does not require nonverbal behavior: textual content alone appears to be a sufficient channel."[40] Such a finding suggests that the spread of emotional contagion does not require face-to-face physical contact. Simply reading the text of a friend is enough to pass the contagion along among those connected in a social network.

The power to influence others through social networks is not limited to our emotions. Christakis and Fowler reveal that not only do we have the ability to influence the weight gain of our friends, friends of friends, and friends of friends of friends—with three degrees of separation—we have the ability to do so from as far away as one thousand miles.[41] In a study of more than five thousand subjects, Christakis and Fowler were able to map social networks to discover that obesity as well as weight loss seemed to happen in

---

[37]Christakis and Fowler, *Connected*; Nicholas A. Christakis and James H. Fowler, "The Spread of Obesity in a Large Social Network over 32 Years," *New England Journal of Medicine* 357 (2007): 370-79; J. Niels Rosenqust, Joanne Murabito, James H. Fowler, and Nicholas Christakis, "The Spread of Alcohol Consumption Behavior in a Large Social Network," *Annals of Internal Medicine* 152 (2011): 426-33; J. Niels Rosenquist, James H. Fowler, and Nicholas Christakis, "Social Network Determinants of Depression," *Molecular Psychiatry* 16 (2011): 273-81; Nicholas A. Christakis and James H. Fowler, "The Collective Dynamics of Smoking in a Large Social Network," *The New England Journal of Medicine* 358 (2008): 2249-58; James H. Fowler and Nicholas A. Christakis, "Dynamic Spread of Happiness in a Large Social Network: Longitudinal Analysis over 20 Years in the Framingham Heart Study," *British Medical Journal* 337 (2008): 1-9.

[38]Adam D. I. Kramer, Jamie E. Guillory, and Jeffrey T. Hancock, "Experimental Evidence of Massive Scale Emotional Contagion Through Social Networks," *Proceedings of the National Academy of Sciences* 111, no. 29 (2014): 8788-90.

[39]Ibid., 8789.

[40]Ibid., 8790.

[41]Christakis and Fowler, *Connected*, 114.

clusters of network connections.[42] One of the particular results they dis-covered was that if one subject became obese, that person's friends were more than twice as likely to become obese themselves.[43] Mutual friendships were no less affected by this weight gain. It was determined that mutual friends were "roughly 20 percent more likely to become obese if the friend of a friend became obese—even if the connecting friend didn't put on a single pound. Indeed, a person's risk of obesity went up about 10 percent even if a friend of a friend of a friend gained weight."[44] They also discovered that other behaviors, including smoking and drinking, were similarly influ-enced by social networks. Taking from the study, Clive Thompson reported that "a friend taking up smoking increased your chance of lighting up by 36 percent, and if you had a three-degrees-removed friend who started smoking, you were 11 percent more likely to do the same."[45]

A network of friends influences others either directly or indirectly. Charles Kadushin underscores this by providing an explanation: "People with like characteristics tend to be connected and . . . connected people tend to have an effect on one another."[46] Marketing experts understand this phe-nomenon and have often used social networks to diffuse goods and services to the general population.[47] Kadushin points to the issue of viral marketing as a way of tapping into one's connections to spread, like a disease, the per-suasion to make certain purchases.

Due to our placement in a social network, we are influencing and being influenced by the degrees of separation we have with one another. Those individuals we mentioned earlier who occupy a place of centrality as a hub in a social network are the most influential and the most vulnerable. They can exert great influence on others, but they also get exposed to more in-fluence from others. Our behaviors, attitudes, values, and even appearance come under the influence of others within our social networks. The in-fluence of relationships has far more impact than we may have previously

---

[42]Clive Thompson, "Are Your Friends Making You Fat?," *New York Times Magazine*, last modified September 9, 2010, www.nytimes.com/2009/09/13/magazine/13contagion-t.html.

[43]Christakis and Fowler, *Connected*, 109.

[44]Thompson, "Are Your Friends Making You Fat?"

[45]Ibid.

[46]Kadushin, *Understanding Social Networks*, 9.

[47]Everett M. Rogers, *Diffusion of Innovation*, 5th ed. (New York: Free Press, 2003).

realized. For that reason, we suggest that the social dimension, more than any other, has contributing value to one's spiritual development and growth. Based on this information, we believe there is likely a correlation between the immersion of the self in healthy relationships and spiritually formative experiences. Conversely, marginalizing those social connections can have a detrimental effect on development.

If growth and development are as significantly affected by social networks as we think they might be, and if spiritual formation follows the same path of development as the rest of the person, it stands to reason that we can measure our own growth and development, as well as others', by assessing the health, location, and number of relationships in our social network. The key component in those relationships is reciprocity. We know from examples and descriptions of social networks that there is a certain sense in which reciprocity or altruistic behavior is akin to the cement that binds those connections together. Social scientists would likely argue that if we are kind to one another, there is a greater likelihood that networks will continue to grow and evolve. Christakis and Fowler underscore this reality by noting that "once networks are established, altruistic acts—from random acts of kindness to cascades of organ donation—can spread through them."[48] Reciprocity then fosters growth among those who are part of a social network and creates a spreading mass of connection points.

Christakis and Fowler point to studies that purport to show altruistic behaviors as more likely to spread than selfish ones. One reason for this is the human tendency to imitate others, and in that process, people in a social networks generally reciprocate. Adoption of an innovation is another form of imitation, which, in the earliest days of social network research, gave rise to "the laws of imitation."[49] Christakis and Fowler quote social critic Eric Hoffer's observation that "when people are free to do as they please, they usually imitate each other."[50] If we reciprocate certain positive behaviors, there is greater likelihood that we will create a vehicle by which we can accomplish and facilitate future plans, ideas, or actions. If we respond in like manner to a kind deed, there is a greater likelihood that we will see

---

[48]Christakis and Fowler, *Connected*, 296.
[49]Rogers, *Diffusion*, 41.
[50]Christakis and Fowler, *Connected*, 112.

reciprocity in return. Additionally, socially networked people create a force of action because of the strength found in unity.

OUR SPIRITUAL NETWORK

Social networks consist of interconnections, interactions, and mutual influence that spread through those social connections. The spiritual network or ecology of the body of Christ partakes of identical elements that produce mutual spiritual influence manifested as mutual spiritual growth. A contagion effect occurs within the body of Christ as individual members grow in their faith because of their common connection to Christ and subsequent connections with one another. James Howard notes that "the community plays a significant and essential role in the progressive sanctification of the believer."[51] Fellowship between Christians takes the connections and interactions we have with one another and makes them conduits for the spiritual contagion of holiness. Whitney reminds us that the New Testament is replete with terms—like *flock*, *body*, *building*, and *household*—that underscore the integration between the individual and the whole and emphasize the communal nature of growing in Christ (see Acts 20:28; 1 Cor 12:12; Eph 2:19, 21).[52] The more reciprocal exchanges we have with one another in the body of Christ, the more opportunities we have to spread the contagion of holiness. The more connected we are to one another through reciprocal interactions, the greater the likelihood for continued growth and development of the whole person. We connect in more ways to more people than we may realize; what we do individually has an effect on others, either directly or indirectly. The connections and interactions that characterize the functioning of the body of Christ, as Paul describes it in 1 Corinthians 12 and Ephesians 4, produces mutual growth of the body.

What we see reflected in the book of Ephesians is similar to the point Jesus makes in the Gospels. Jesus used parables to communicate the relationship between kingdom growth and how things grow in nature. He tells us to examine the lilies and how they grow if we want to know something

---

[51]James M. Howard, *Paul, the Community, and Progressive Sanctification: An Exploration into Community-Based Transformation Within Pauline Theology* (New York: Peter Lang, 2007), 84.

[52]Donald S. Whitney, *Spiritual Disciplines for the Christian Life* (Colorado Springs: NavPress, 1994).

about growth in the kingdom of God. One thing we know about lilies is that they grow in groups. Only very rarely might one find a lone plant growing in an environment not conducive to lilies. Normal growth results from being in the right environment, and a good environment contains the connection to other dimensions found within the ecology of the person. There is a holistic trajectory to the way we grow and develop both personally and corporately. Jesus was a perfect example of whole-person development (Eph 4:13). Luke, the Gospel writer, says Jesus grew in wisdom, stature, and favor with God and with people (Lk 2:52). The human hand is an apt illustration of this personal ecology of whole-person transformation into the "fullness of Christ" because it represents the way all of the human dimensions of the person work together to form the whole. Each of the digits represents five dimensions of human development, while the palm represents the spiritual component of our formation. It would be illogical to think of a normally functioning, healthy hand without fingers. Similarly, we would consider a hand with no palm and only digits abnormal. Each of these components of the hand acts in an interrelated way. The order of the way we are created suggests an interactive dynamic, rather than each aspect acting independently of the others. We might say that each one of us contains a perfect illustration of whole-person ecology in our hand. Figure 3 illustrates our personal ecology composed of six connected and interacting dimensions designed to instigate whole-person transformation.

Figure 3. Six developmental dimensions

SUMMARY

Our place within social networks and the relationships we develop with others lends itself to mutual, whole-person development. Reciprocity in relationships doesn't simply enhance social development; in enhances physical, psychological, spiritual, emotional, and moral development as well. The absence of social connections with others leads to adverse outcomes. In his book, *A Cry Unheard,* James Lynch provides scientific evidence that high blood pressure, heart disease, cancer and other diseases are more prevalent and more deadly among those living alone than among those who are married or living with family and friends.[53] A number of studies show an increase in the rate of the "widowhood effect," that is, higher rates of mortality following the death of a spouse. J. Robin Moon et al. conducted a study on the widowhood effect among people aged fifty or older who outlived their spouses. They found that there is a 66% chance the surviving spouse will pass away within the first three months after the death of their spouse.[54] The only way to overcome this negative effect is for surviving spouses to stay connected with family, friends, and neighbors, referred to by researchers as "social participation."[55] Researchers define social participation as "social interactions with persons other than a spouse" and includes formal "meeting attendance, religious participation, and volunteer obligations" as well as informal "telephone contact and social interactions with friends."[56]

Social networks explain how it is possible for people, connected in some sort of social bond or relationship and engaged in a variety of reciprocal interactions, to influence each other mutually for good or for ill. In this group of chapters (part two), the hidden power of learning, digital, and social ecologies have alerted us to the importance of thinking ecologically not just about spiritual formation but other aspects of our lives as well. Our high-altitude view helps us see just how interconnected we are to other people and how this affects how we learn, relate to others, and navigate an increasingly networked society. All of the ecological environments we inhabit as humans partake of various kinds of connections (interpersonal, social, educational, societal, etc.) and create a number of

---

[53]James J. Lynch, *A Cry Unheard: New Insights into the Medical Consequences of Loneliness* (Baltimore: Bancroft Press, 2000).

[54]J. Robin Moon et al., "Short- and Long-Term Associations Between Widowhood and Mortality in the United States: Longitudinal Analyses," *Journal of Public Health* 36, no. 3 (2014): 382-89.

[55]Rebecca L. Utz, Deborah Carr, Randolph Nesse, and Camille B. Wortman, "The Effect of Widowhood on Older Adults' Social Participation: An Evaluation of Activity, Disengagement, and Continuity Theories," *Gerontologist* 42, no. 4 (2002): 522-33.

[56]Ibid., 522.

interactive exchanges through which we mutually influence one another positively or negatively.

In part three, we will now explore in some detail how the apostle Paul used similar ecological constructs to identify how Christians connect spiritually to Christ and interact with one another to produce a mutually beneficial spiritual outcome. A key to understanding spiritual growth is to recognize the necessity of our spiritual connection to Christ and our spiritual interconnections to one another. Paul had a specialized vocabulary to describe both these connections and the interactions that flow between them. We now turn to this often-neglected feature of Pauline theology.

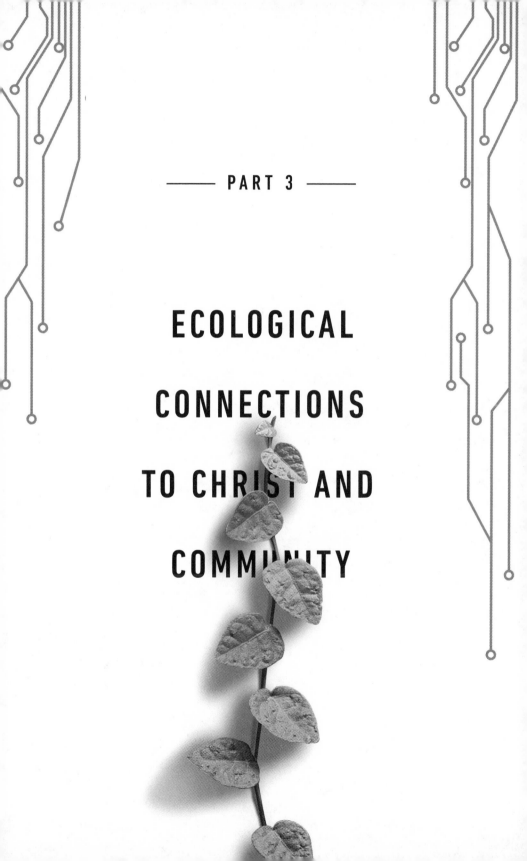

— PART 3 —

ECOLOGICAL

CONNECTIONS

TO CHRIST AND

COMMUNITY

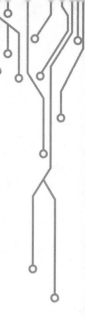

── NINE ──

ECOLOGICAL CONNECTION
TO CHRIST

Join'd to Him, we then shall shine all immortal, all divine!

CHARLES WESLEY, "CHRIST THE LORD IS RISEN TODAY"

The apostle Paul, more than any other New Testament writer, provides us with the concepts, vocabulary, and theology of spiritual formation. Through the many letters he wrote both to churches he founded and those he had yet to visit, we find the riches of his understanding for what it meant to be a disciple of Jesus. Paul's instruction and insight are intriguing in how compatible they are with the ecological model of spiritual formation we have pursued. While Pauline scholars and others have rightly stressed the significance of our "union with Christ" as central to Paul's theology, many have overlooked the fact that Paul also uses identical language to refer to our union with other Christians who are members of the body of Christ. Since all growth in God's universe is ecological growth enabled by connections to and interactions with other organisms, we should not be surprised to find those same features present in Paul's understanding of the spiritual ecology of the church as the body of Christ.

We have woods all around our home and we enjoy not only looking at them but walking through them. When we look at trees or walk through the woods, we only focus on what we see above ground. Forest ecologists like Dr. Suzanne Simard and others, however, have drawn our attention to the underground network that lies beneath the surface. She and others have discovered a vast underground interlocking fungal and root network that connects all of the trees together. Through these vast, hidden networks the trees share nutrients, exchange carbon, and provide life-giving water to each other in a highly mutualistic interchange. Simard et al. discovered that "plants within communities can be interconnected and exchange resources through a common . . . network."[1] When we observe the beauty and breadth of the body of Christ we can miss the hidden connections and interactions that function like the underground network Dr. Simard describes.

We find in Paul's descriptions of the spiritual ecology of the body of Christ similar spiritual connections and interactions that provide the necessary spiritual nutrients for growth. The starting point for seeing these fundamental components of an ecosystem is Paul's use of the Greek prefix *syn*, meaning "together with." It is Paul's distinctive prefix, combined with other verbs and nouns to depict our connection to Christ in all of its multiple applications. Murray Harris defines the preposition *syn* as "together with" or "jointly with," signifying the idea of close association and accompaniment.[2] Brendan McGrath expands our understanding of the meaning of the preposition by adding that Paul used it "to express as forcibly as possible the intimacy of the connection of the Christian with Christ."[3] This is why we prefer to use the language of "connection with Christ" to express the force of the *syn* prefix. Ernest Best uses such language in his attempt to describe exactly what Paul meant by his "with Christ" language: "The believer is never said to be Christ . . . but he is connected to Christ by a bond similar to that which links two parts of an organism."[4] Connection terminology is the language of ecology that emphasizes the way in which different components of an ecosystem organically connect to each other to form a whole of interconnecting organisms that share nutrients and resources. This is why Barry Commoner, in his four basic laws of ecology, gives "everything is connected to everything else" as the first law.[5] He explains that this law "reflects the existence of the elaborate network of interconnections in the ecosphere . . . that consists of multiple interconnected parts, which act on one another."[6]

---

[1]Suzanne W. Simard et al., "Net Transfer of Carbon Between Ectomycorrhizal Tree Species in the Field," *Nature* 388 (1997): 580.

[2]Murray J. Harris, *Prepositions and Theology in the Greek New Testament* (Grand Rapids: Zondervan, 2012), 199.

[3]Brendan McGrath, "'Syn' Words in Saint Paul," *Catholic Biblical Quarterly* 14 (July 1952): 219-26.

[4]Ernest Best, *One Body in Christ* (London: SPCK, 1955), 55.

[5]Barry Commoner, *The Closing Circle: Nature, Man, and Technology* (New York: Alfred A. Knopf, 1971), 16.

[6]Ibid.

THE NEED FOR CONNECTION WITH CHRIST

Before we explore the significance of Paul's *syn*-connection language, it is important to set the context that established the need for Paul to use this language and to see how such language fits into Paul's larger biblical landscape.[7] As C. E. B. Cranfield and James Dunn explain, the backdrop on the stage of Paul's theology is Genesis 1–3.[8] Part of this backdrop includes the disruption caused by human sin upon the spiritual fellowship enjoyed in the perfect ecology of the Garden of Eden.

Once human sin disconnected us from perfect fellowship with God in the garden, all harmony, peace, and wholeness no longer existed, and in its place, we find spiritual alienation (spiritual disconnection), divine hostility, and social fragmentation. Paul says of Jesus in Colossians 1:21-22, "And although you were formerly alienated and hostile in mind, *engaged* in evil deeds, yet He [Christ] has now reconciled you in His fleshly body through death, in order to present you before Him holy and blameless and beyond reproach." The "former alienation" to which Paul speaks here is what the Genesis narrative refers to when it describes God removing Adam and Eve from the garden, thus ending the previously enjoyed fellowship. Humanity disconnected (alienated) the connection (fellowship) we enjoyed with God in the Garden of Eden through sin, and we must be reconnected (reconciled) by God through the death of Christ on the cross. The need for reconnection, as a result of human sin, reflects the personal nature of the original connection between humans and God and the personal nature of the disconnection. The work of Christ on behalf of all humanity is not simply a legal proceeding that balances the scales of divine justice. Although it certainly is that, it is also a personal matter motivated by divine love (Jn 3:16).[9]

---

[7]Grant Macaskill, *Union with Christ in the New Testament* (Oxford: Oxford University Press, 2013), rightly cautions against a purely grammatical interpretation of the significance of the *syn*-compound language in Paul and elsewhere in the New Testament. He states, "Instead, we must read such grammatical constructions in relation to their contexts, sensitive to the way in which they may be informed by underlying narratives" (219). We agree and seek to contextualize Paul's use of this language, especially within the narrative structure of Ephesians and Philippians.

[8]C. E. B. Cranfield, *A Critical and Exegetical Commentary on the Epistle to the Romans*, vol. 1 (Edinburgh: T&T Clark, 1975); James D. G. Dunn, *The Theology of Paul the Apostle* (Grand Rapids: Eerdmans, 1998).

[9]See N. T. Wright, *Paul and the Faithfulness of God* (Minneapolis: Fortress, 2013), 1038-42, where he discusses the relationship between "juridical" and "participationist" categories.

The result of our reconnection to God through Christ is union with Christ in which God credits all the work of Christ to our account. Our union with Christ provides a vital and necessary connection to all the required spiritual nutrients Christians need to "grow in the grace and knowledge of our Lord and Savior Jesus Christ" (2 Pet 3:18). When Jesus described our union with him in John 15, he explained the relationship in organic ecological word pictures as the relationship between vine and branches in a vineyard. Hans Burger describes John's approach to our connection with Christ as "reciprocal inhabitation."[10] More precisely, John is representing the view that Jesus expounded in his instructions to his disciples since the language and images originate with Jesus. Nevertheless, his description of "reciprocal inhabitation" is accurate when situated within the ecology of the vineyard. The reciprocal nature of the connections and interactions between vine and branches conveys the spiritual nutrients needed for the branches to produce fruit.

SIMPLE PREPOSITION *SYN*

Occasionally in the epistles, Paul uses the preposition *syn*, meaning "together with" or "connected to," with a pronoun referring to Christ (Rom 6:4, 5, 6, 8; 8:32; 2 Cor 13:4; Eph 2:6; Col 2:12, 13; 3:4; 1 Thess 4:14; 5:10; 2 Tim 2:11, 12). Paul uses this form to stress our connection *to the person of Christ*. Romans 8:32 is a representative sample of the other occurrences listed above: "He who did not spare His own Son, but delivered Him over for us all, how will He not also with Him (*syn autō*) freely give us all things?" Paul links the believer's reception of "all things" God has to offer to the believer's connection to Christ, represented by the preposition *syn* paired with the personal pronoun. This statement comes at the end of a series of *syn autō*, meaning "with Him," phrases earlier in Romans that specify the "things" Paul alludes to in Romans 8:32 (Rom 6:4-8). The believer's connection to Christ in salvation provides "all . . . things" the Christian needs and puts the Christian in an invincible position (Rom 8:37-39).

---

[10]Hans Burger, *Being in Christ: A Biblical and Systematic Investigation in a Reformed Perspective* (Eugene, OR: Wipf and Stock, 2009), 280.

SYN CHRISTŌ

The simple preposition *syn* coupled with a personal pronoun for Jesus is a complement to the more decisive use of the *syn Christō* language that specifies, more precisely, with whom Christians connect.[11] According to John Harvey there are twenty-one *syn*-compounds that "speak of the believer's association with Christ" and cover all of the various forms we will consider here.[12] There are four New Testament occurrences of *syn Christō* (Rom 6:8; Col 2:20; 3:3; Phil 1:23), one instance of *syn Iēsou* ("with Jesus," 2 Cor 4:14), and one instance of *syn Kyriō* ("with the Lord," 1 Thess 4:17). When Paul uses such phrases in combination with *syn*, they form what Robert Tannehill calls a "motif," and they express a fundamental truth about the relationship between the Christian and Christ.[13] This fundamental truth places the "emphasis on the realism of the union between the individual Christian and Christ."[14] The bond created by faith in the death of Christ on the cross that "cleanse[s] us from all unrighteousness" connects every believing Christian directly and intimately with the person of Christ (1 Jn 1:9). This connection is so strong in Paul's mind that it is impossible to separate or distinguish between the risen Christ and those who are "with" him. Christ revealed this foundational truth to Paul on the road to Damascus recorded in Acts 9:4-5: "'Saul, Saul, why are you persecuting Me?' . . . 'I am Jesus whom you are persecuting.'" Although Luke tells us Saul was "breathing threats and murder against the disciples of the Lord," Jesus informs Saul that to persecute his followers is the same as persecuting him, the risen Lord (Acts 9:1). Whereas John A. T. Robinson sees the connection between Christ and his body as literal and physical, Robert Gundry reads the same texts and concludes that the connection is figurative.[15] As Best cautions, we cannot infuse one aspect

---

[11] Although it would be proper to include Paul's use of *en Christō* ("in Christ") within a discussion of our union with Christ, our focus here is exclusively on the *syn* preposition and *syn*-compounds. For a full treatment of Paul's *en Christō* usage, see Dunn, *Theology*, 396-401; Herman Ridderbos, *Paul: An Outline of His Theology* (Grand Rapids: Eerdmans, 1975), 57-64; Udo Schnelle, *Apostle Paul: His Life and Theology* (Grand Rapids: Baker Academic, 2003), 481-82.

[12] John D. Harvey, "The 'With Christ' Motif in Paul's Thought," *Journal of the Evangelical Theological Society* 35, no. 3 (September 1992): 331.

[13] Robert C. Tannehill, *Dying and Rising with Christ* (Eugene, OR: Wipf and Stock, 1967), 6.

[14] Barnabas M. Ahern, "Christian's Union with the Body of Christ in Cor, Gal, and Rom," *Catholic Biblical Quarterly* 23, no. 2 (1961): 209.

[15] John A. T. Robinson, *The Body: A Study in Pauline Theology* (London: SCM Press, 1952); Robert H. Gundry, *Sōma in Biblical Theology: With Emphasis on Pauline Anthropology* (Grand Rapids: Academie Books, 1976).

of Paul's theology into all the other intricate aspects of it. While we can certainly say, "The Church is in Christ . . . Yet on the other hand there is much
that distinguishes Christ from the church; he is not only united to it but
stands over against it as its Redeemer and Lord."[16] We do not think that we
have yet any satisfactory solution to how all of these various aspects of Paul's
theology integrate into a synthetic whole. There remains a certain degree of
mystery and incompleteness to our present understanding that we must
accept even though it may create a measure of intellectual dissatisfaction.

The connection we have with Christ is so enduring that Paul speaks of it
as still existing even in death and in life beyond death: "But I am hard-
pressed from both *directions*, having the desire to depart and be with Christ
(*syn Christō*), for *that* is very much better" (Phil 1:23). In Paul's mind, our
connection to Christ spans the scope of our past union with Christ and
looks forward to our eternal union with him.

VERTICAL *SYN*-COMPOUNDS

Through the *syn-Christō* language, Paul stresses our connection to *the person
of Christ*. However, through the vertical *syn*-compound language, Paul emphasizes our connection to *the work of Christ*. Murray Harris points out that
"there are more New Testament compounds formed with *syn* than with any
other preposition."[17] The abundant number of them that appear in Paul's
epistles—85 distinct compounds—appear 158 times.[18] We distinguish between those *syn*-compounds that direct our attention to the Christian's connection to Christ in salvation (vertical) and those that focus on the connections among Christians (horizontal), which unite them in "fellowship of the
Spirit" (Phil 2:1). Here we focus on the vertical *syn*-compounds. The *syn*-
compounds get their name because the preposition *syn* forms a compound
or hybrid word with a noun or verb.

Brendan McGrath offers a helpful analysis of Paul's use of the *syn*-com-
pounds and remarks that Paul "possesses the ability to take a quite ordinary
word, and, understanding it in the most strictly literal sense, cram into it

---

[16]Best, *One Body*, 195.
[17]Murray J. Harris, *Prepositions and Theology in the Greek New Testament* (Grand Rapids: Zondervan, 2012), 199.
[18]Harvey, "'With Christ' Motif," 331.

more meaning and a deeper significance than Plato could have dreamed possible."[19] Paul packs into these distinctively hybrid words deep theological meaning regarding the intimate bond created between the Christian and Christ at the moment of conception—at our spiritual rebirth. Although sprinkled throughout the epistles traditionally ascribed to Paul, the *syn*-compounds cluster in two epistles: Romans and Ephesians. Paul's use of the *syn*-compound in Romans focuses primarily but not exclusively on the vertical union between Christians and Christ. Conversely, Paul's use of the *syn*-compound in Ephesians focuses primarily but not exclusively on the horizontal relationship between Christians—especially the Jew-Gentile relationship.

In the epistle to the Romans, Paul draws our attention to the "death-resurrection motif in the compounds uniting the believer to Christ."[20] Dunn apprehends what one might call the shockwaves of salvation that emanate from the decisive events of the death and resurrection of Jesus. The force of the language compels one to recognize "a quite profound sense of participation with others in a great cosmic movement of God centered on Christ and effected through his Spirit."[21]

While space does not permit a full treatment of all the redemptive or vertical *syn*-compounds, we include a few that represent their usage in the Pauline epistles traditionally attributed to Paul and follow the order of salvation.

Galatians 2:20 (see also Rom 6:6): "I have been crucified with (systauroō) Christ." The vertical *syn*-compound Paul uses here and in Romans 6 is identical to the phrase used by Matthew, Mark, and John to describe the thieves who were "crucified with Him" (Mt 27:44; Mk 15:32; Jn 19:32). Paul takes the literal use of the term from the Gospels and applies it figuratively to our relationship with Jesus. In Romans 6:6, he specifies that the "old self" was crucified with Jesus on the cross. The dramatic force of this language marks the decisive manner with which the death of Jesus stripped sin of its power over us. Paul describes the effect of the cocrucifixion in Colossians 2:14 when he writes, "Having canceled out the

---

[19]Brendan McGrath, "The Doctrine of Christian Solidarity in the Epistles of Saint Paul" (doctor of sacred theology diss., University of Ottawa, 1952), 406.

[20]Dunn, *Theology*, 403.

[21]Ibid., 404.

certificate of debt . . . against us . . . having nailed it to the cross." Through our cocrucifixion with Christ, the debt of sin is paid, and sin no longer has any authority or power over us. Since death "has lost its sting" over Jesus, demonstrated by his resurrection from the dead, and since Christians were cocrucified with Christ, it no longer has any "STING" for those who are connected to Christ in his death (1 Cor 15:55-56). This is why Paul comforts the believers at Thessalonica not to grieve the loss of those who have died in Christ "as do the rest who have no hope" (1 Thess 4:13).

Romans 6:4: "Therefore we have been buried with (synthaptō) Him through baptism into death." The second aspect of the work of Christ with which we are connected follows upon our cocrucifixion with Christ; we are coburied with Christ. The burial of Jesus confirmed the reality of his death by crucifixion. We find the illustration of this burial with Christ symbolized in water baptism. While other forms of baptism (sprinkling or pouring) are legitimate expressions of this burial with Christ, baptism by immersion certainly illustrates it more dramatically. Even Martin Luther, who practiced infant baptism, in 1519 with "The Holy and Blessed Sacrament of Baptism" recognized the tradition of immersing infants into the baptismal font because it accorded with the true meaning of *baptismos*: "to plunge something completely into the water, so that the water covers it."[22]

Ephesians 2:5-6: "Made us alive together with (syzōopoieō) Christ . . . and raised us up with (synegeirō) Him."[23] Paul uses the correct order when linking believers to the work of Christ. We were first "made alive together with Christ" and then we were "raised up with Him" from death. This is the "victory through our Lord Jesus Christ" to which Paul alludes in 1 Corinthians 15:57. The Christian thus stands as Christ does in a position of being victorious over the "last enemy" (1 Cor 15:26).

Romans 8:17: "So that we may also be glorified together with (syndoxazō) Him." Paul contrasts the believer's present experience, in which we "suffer with *Him*" (*sympaschō*), to that time in the future when we will also "be glorified together with *Him*." The glorification of Christ

---

[22]Martin Luther, "The Holy and Blessed Sacrament of Baptism" (H. E. Jacobs, translator), paragraph 1.

[23]Because of the peculiar phonology of Koine Greek, the lexical form of the preposition *syn* (*sun*) changes when the Greek letter *nu* ("n") combines with other consonants resulting in different spellings of the preposition such as *sug*, *sul*, *sum*, *sus*, and *suz*.

involves his ascension and exaltation to the right hand of the Father, where he "SAT DOWN" upon the completion of his saving work to which Paul connects us through the various *syn*-compounds (Lk 24:51; Acts 1:11; Phil 2:9-11; Heb 10:12).

Ephesians 2:6: "And seated us with (sygkathizō) Him in the heavenly places *in Christ Jesus*." When Christ was exalted to the Father and "sat down," Paul says Christians were "seated together with Him" and therefore now enjoy the privileged position of their glorified Lord. God takes this action toward believers based upon our connection with (*syn*) Christ (Eph 2:4). Harold Hoehner makes this point when he writes, "It is our union with Christ that gives us the right to be in the heavenly places."[24]

Romans 8:17: "And if children, heirs also, heirs of God and fellow heirs with (sygklēronomos) Christ." The result of being seated together with Christ is that we become fellow heirs with Christ and enjoy the inheritance that God gives his Son and the Son shares with us (Eph 1:11, 14, 18). Of course, Paul's concept of inheritance has its origin in God's covenant with Israel and the promised inheritance of the land. However, even in the Old Testament the promise of land first enunciated in Genesis 12 simply particularizes the view of the entire earth encompassed in Genesis 1–11. This seems to be Paul's perspective in Romans 8:17 as he anticipates the time when Christ's inheritance as God's Messiah will be a renewed and revitalized earth. When the "sons of God" are "revealed," the groaning we experience and the groaning of the earth will be "set free" (Rom 8:18-21). Although Paul specifies the inheritance in Galatians 5:21 as "the kingdom of God," all of his previous references to inheritance in the epistle refer to the inheritance of the land of promise (Gal 3:18, 29; 4:1, 7, 30). Thus N. T. Wright can conclude, "The final 'kingdom of God' is *the whole world, rescued at last from corruption and decay* and living under the sovereign rule of God, exercised through the Messiah's people."[25] Therefore, our inheritance is not some gold-plated heaven in the clouds but a renewed heavens and earth that we will finally enjoy as God's perfect ecology for human habitation.[26]

---

[24]Harold W. Hoehner, *Ephesians: An Exegetical Commentary* (Grand Rapids: Baker Academic, 2002), 335.

[25]Wright, *Paul*, 1:367. Italics his.

[26]See Paul Enns, *Heaven Revealed* (Chicago: Moody Press, 2011).

INDICATIVE AND IMPERATIVE

Every Christian's connection with the person and work of Christ, illustrated in the vertical *syn*-compounds and grounded in our *syn Christō* relationship, forms an indissoluble organic bond that forms the cornerstone of a spiritual ecology from which we grow together toward full maturity (Eph 2:19-21; Gal 6:6-10). We see Paul teaching this most clearly in Romans 6:5: "For if we have become united with *Him* in the likeness of his death, certainly we shall also be [united] *in the likeness* of his resurrection." The *syn*-compound here is *symphytos* ("grow with"), and it is obviously a term drawn from the bio-ecological realm of creation. It is the term for living things that organically connect together, where two shoots grow together to form a single plant or like the relationship between the head and the rest of the body. Here again, Paul calls upon a bio-ecological metaphor to help us understand how believers connect with Christ. We could say, based upon his usage of this *syn*-compound, that Paul understands the connection between Christ and Christians in ecological categories. This ecological or organic connection unites us in a profound way with Christ and his redemptive work on our behalf. The imagery Paul uses in Romans 6:5, with *symphytos*, is very similar to the imagery Jesus uses in a different context between the vine and branches. As Best writes, "The believer . . . is connected to Christ by a bond similar to that which links together two parts of an organism."[27]

The connection with Christ encompasses all aspects of his redemptive work on our behalf from death to exaltation. Vincent Taylor summarizes this nicely, writing,

> When Christ died something happened once for all, not only to him who died, but all for whom he died. They also died with him upon the cross. It is as if believers were with Christ upon the cross; when he died, they died; when he was buried, they were buried; when he rose, they rose.[28]

The indicative truth is that the Christian's "old self was crucified with *Him* [Christ]" on the cross (Rom 6:6). The imperative derived from this is that "even so consider yourselves to be dead to sin" (Rom 6:11). The indicative truth is that Christians were "raised up with Christ" (Col 3:1). The imperative

---

[27]Best, *One Body*, 55.
[28]Vincent Taylor, *Forgiveness and Reconciliation*, 2nd ed. (London: St. Martin's Press, 1952), 119.

truth is "walk in newness of life" (Rom 6:4). The indicative truth is that Christians are "seated . . . with Him [Christ]" (Eph 2:6). The imperative truth is "keep seeking the things above, where Christ is, seated at the right hand of God," and "set your mind on the things above" (Col 3:1-2). The movement from indicative—what is true about our relationship to Christ—to imperative—how we live out that truth—requires the empowerment of the Holy Spirit triggered by "obedience of faith" (Rom 1:5; 16:26). In Philippians 2:12, Paul links faithful obedience to how we "work out [our own] salvation with fear and trembling." The imperatives in Paul's epistles and elsewhere in the New Testament have no meaning if we think of the Christian's role as passive when it comes to our sanctification or spiritual maturity. While we are recipients of the indicative realities Paul describes, we are activists toward the imperatives that flow from those indicatives. The obedience of faith refers to our initial response to the gospel and our subsequent persistence in gospel living. Although there are various interpretations of the grammar of the phrase "*the* obedience of faith" (*hypakoēn pisteōs*) that determine how the phrase should be translated and interpreted, D. B. Garlington marshals considerable evidence for understanding this distinctively Pauline phrase as "obedience which is the product of faith" or "faith's obedience."[29] This leads us to conclude, as Morna Hooker does, that "faith is certainly not to be understood as a form of human works!"[30] Faith derives from our connected relationship to Christ, not from human effort. The spiritual ecology of connected relationships to Jesus and other Christians forms a God-created and Spirit-empowered environment in which faith expresses itself through obedience.

As stated in passing above, the movement from indicative to imperative requires the empowering work of the Holy Spirit in the Christian and in the Christian community. Paul makes this point with the believers in the church at Ephesus when he prays for them at the conclusion of the indicative portion of his epistle (Eph 3:14-21). The hinge that moves the Ephesian church from a mere apprehension of the truths Paul espoused in Ephesians 1–3 to the implementation of those truths in Ephesians 4–6 is the central petition of his prayer for them: "That He would grant you, according to the

---

[29]D. B. Garlington, "The Obedience of Faith in the Letter to the Romans," *Westminster Theological Journal* 52 (1990): 224.

[30]Morna D. Hooker, *From Adam to Christ: Essays on Paul* (Eugene, OR: Wipf and Stock, 1990), 185.

riches of His glory, to be strengthened with power through His Spirit in the inner man" (Eph 3:16). The phrase "through His Spirit" is a prepositional phrase of agency. The Ephesian believers could only be strengthened with power through the agency or support of the Holy Spirit. All the imperatives that follow in Ephesians 4–6 (grow up, be renewed, put on the new self, be kind to one another, forgive each other, etc.) can only be implemented through the power provided by the work of the Holy Spirit (Eph 3:20). However, the triggering mechanism for this empowerment must be the obedient act of performing the imperatives as the apostle admonished them to do.

SIGNIFICANCE FOR ONLINE LEARNING

The *syn* connection we have with Christ supplies all of the spiritual nutrients needed for proper growth and, indeed, gives life that makes possible our growth. Our *syn*-connection to Christ links us to him so securely that nothing "will be able to separate us from the love of God, which is in Christ Jesus our Lord" (Rom 8:39; see also Rom 6:5, 6, 8; 7:4; 8:17). While God has made this great provision for us *syn Christō*, he expects us to activate this connection—animated through the work of the Holy Spirit—by engaging in reciprocal interactions with other connected believers that comprise the spiritual ecology of the body of Christ.

These reciprocal interactions can occur in any environment where Christians gather together (*synerchomai*), whether in a sanctuary for preaching and worship, in a coffee shop for prayer and fellowship, or in an online classroom to study theology. An ecological vantage point offers us a wider field of application as to where and when these encounters take place. Since the connections and interactions we have are part of a spiritual ecology unbounded by space and time, we may enjoy them and benefit from them in and through the digital connections and interactions that increasingly occupy our lives. An ecological perspective also allows us to think more broadly regarding how we engage and interact with other believers. Academic environments, whether on campus or online, where Christians form physical and digital communities, may be fertile environments for the kind of synergistic transactions Paul envisions through his use of the *syn*-compounds. Furthermore, acknowledging the reality of our union with Christ among like-minded learners activates a truth students can take both into their local communities and into their other digital relationships. While this chapter focused attention on the Christian's vertical connection to Christ, in our next chapter we examine our spiritual connections to one another through a study of Paul's use of horizontal *syn*-compounds and their implications for the growth of the body and the growth of students in digital ecologies of learning.

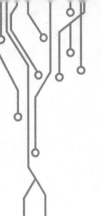

— TEN —

ECOLOGICAL CONNECTIONS
TO CHRISTIANS

[Christ] works on us in all sorts of ways. . . . But above all,
He works on us through each other.

C. S. Lewis, *Mere Christianity*

In the last chapter, we saw Paul using a distinctive *syn*-compound vocabulary to explain how every believer and the church in toto connects to Christ. Theologians refer to this connection as our "union with Christ" or "dying and rising with Christ," and we have this union in respect to every facet of Christ's work of salvation: crucifixion, death, burial, resurrection, and ascension (see Rom 6:3-5).[1] We refer to this distinctive collection as "vertical *syn*-compounds" because they focus exclusively on our vertical connection to Christ as the head of the body.

There is a great deal of scholarly attention given to the vertical *syn*-compounds, which explain the Christian's union with Christ.[2] However, Paul used this same *syn*-compound construction to describe the spiritual connections Christians have with other members of the body of Christ. We refer to this unique set as "horizontal *syn*-compounds," which appear in Paul's epistles because they focus exclusively on our interpersonal relationships

---

[1]Robert C. Tannehill, *Dying and Rising with Christ: A Study in Pauline Theology* (Eugene, OR: Wipf and Stock, 2006); Robert Letham, *Union with Christ in Scripture, History, and Theology* (Philadelphia: P&R, 2011).

[2]Grant Macaskill, *Union with Christ in the New Testament* (Oxford: Oxford University Press, 2013); Constantine R. Campbell, *Paul and Union with Christ: An Exegetical and Theological Study* (Grand Rapids: Zondervan, 2012); J. Todd Billings, *Union with Christ: Reframing Theology and Ministry for the Church* (Grand Rapids: Baker Academic, 2011).

with other Christians in the body of Christ. James Dunn counts twenty-seven distinct *syn*-compounds that occur fifty-four times, and Sang-Won Son counts twenty-seven distinct *syn*-compounds that occur fifty-six times.[3] While several of Paul's epistles would qualify for comment, we focus our attention on Paul's epistle to the Philippians since it contains more *syn*-compounds than any other epistle traditionally attributed to Paul.

HORIZONTAL *SYN*-COMPOUNDS IN PHILIPPIANS

In Philippians, these horizontal *syn*-compounds appear in a discussion of disunity in a local assembly. Paul's horizontal *syn*-compound usage in Philippians provides the hermeneutical rationale for extrapolating beyond the Jew/Gentile relationship in his other epistles. In Philippians, one of the major issues Paul confronts is a general disunity among believers without regard to ethnic identity. Raymond Brown identifies three components to the situation as it existed at the time Paul wrote the letter: (1) internal dissension, (2) external opposition, and (3) infiltrating adversaries.[4] As Gordon Fee suggests, all of these situational components are interrelated and form a unique situational matrix Paul addresses with the "special vocabulary of Philippians."[5] However, what is missing from Fee's otherwise helpful analysis of Paul's vocabulary is any reference to the significant number of *syn*-compounds and how this usage relates to the situational matrix Paul confronts in his letter to the church. In fact, J. Paul Sampley points out that "more words per page are formulated with the *syn* [together] prefix than in any other Pauline letter."[6] Lincoln makes the same observation about the frequency of the *syn*-compound in Philippians and concludes that "it would be hard to imagine how he could make his point . . . any more forcefully."[7] Joseph Marchal observes particularly that "*syn* appears

---

[3]James D. G. Dunn, *The Theology of Paul the Apostle* (Grand Rapids: Eerdmans, 1998); Sang-Won Son, *Corporate Elements in Pauline Anthropology: A Study of Selected Terms, Idioms, and Concepts in the Light of Paul's Usage and Background* (Rome: Editrice Pontificio Istituto Biblico, 2001).

[4]Raymond E. Brown, *An Introduction to the New Testament* (New Haven: Yale University Press, 1997).

[5]Gordon D. Fee, *Paul's Letter to the Philippians* (Grand Rapids: Eerdmans, 1995), 18.

[6]J. Paul Sampley, "Reasoning from the Horizons of Paul's Thought World: A Comparison of Galatians and Philippians," in *Theology and Ethics in Paul and His Interpreters: Essays in Honor of Victor Paul Furnish*, ed. Eugene H. Lovering Jr. and Jerry L. Sumney (Nashville: Abingdon, 1996), 121.

[7]Andrew T. Lincoln, "Communion: Some Pauline Foundations," *Eccessiology* 5 (2009): 154.

as a prefix or preposition twenty times."[8] Seventeen of these *syn* occurrences
are *syn*-compounds.[9] If we include other "special vocabulary of Philippians"
with the frequent *syn*-compounds, what emerges is a motif of connectedness
relating to the situational matrix Paul addresses in Philippi. Included in such
a motif is the *syn* preposition and prefix already referenced, *koinōnia* terms,
and occurrences where Paul compounds *syn* with *koinōnia*.[10] Summarizing,
we might say Paul emphasizes the connection in Christ between himself and
"all" the Philippians to rectify the emerging problem of dissension before it
reaches the level Paul confronted in Corinth (Phil 1:1, 4, 7, 8; 1 Cor 1:10-13; 3:1-4;
11:17-19). Further, if they, as a church, are going to withstand the external opposition (political and theological) they must stand in solidarity with one another. The motif of connectedness encompasses various kinds of connections
that Paul wished to emphasize with the church for whom he had such affec-
tionate feelings (Phil 1:7; 2:17-18; 4:15-16). He highlights the following connections: (1) between the believer and Christ (Phil 3:10, 21); (2) between the believer and the Holy Spirit (Phil 2:1); (3) between Paul and the church (Phil 1:4,
7, 25, 27; 2:17; 3:17; 4:14, 15); (4) between the Philippian believers in the church
(Phil 2:2, 3; 4:2); and (5) between Paul and his coworkers (Phil 2:25; 4:3). These
connections are applied in the broadest possible way to the entire congregation
as a local body of believers through his use of the "all" language sprinkled
throughout the epistle. This broader motif of connectedness subsumes themes
previously observed in Philippians, such as friendship, partnership, and unity.[11]
All of these themes reflect different aspects of a more general theme of connectedness instigated by a combination of internal and external factors that
threatened the cohesion of this community.

As William Dalton affirms, "If a regular pattern of words and ideas is
repeated in a way which reveals the inner movement and meaning of the
text," we have evidence of a cohesive theme or motif that unifies the entire
letter.[12] For our purposes, we are interested only in the frequency of the *syn*

[8]Joseph A. Marchal, *Hierarchy, Unity, and Imitation: A Feminist Rhetorical Analysis of Power Dynamics in Paul's Letter to the Philippians* (Atlanta: Society of Biblical Literature, 2006), 26n17.
[9]Phil 1:7, 22, 23, 25, 27; 2:2, 17, 18, 25a, 25b; 3:10, 17, 21; 4:3a, 3b, 3c, 3d, 14.
[10]*Syn*: Phil 1:1, 7, 22, 23a, 23b, 27; 2:2, 17, 18, 25a, 25b; 3:10, 17, 21; 4:3a, 3b, 3c, 3d, 14, 21. *Koinōnia*:
Phil 1:5; 2:1; 3:10; 4:15. Compound *syn* and *koinōnia*: Phil 1:7, 4:14.
[11]See Fee, *Philippians*, 1995; J. Paul Sampley, *Pauline Partnership in Christ* (Philadelphia: Fortress,
1980); J. B. Lightfoot, *St. Paul's Epistle to the Philippians* (Grand Rapids: Zondervan, 1973); G.
Walter Hansen, *The Letter to the Philippians* (Grand Rapids: Eerdmans, 2009).
[12]William J. Dalton, "The Integrity of Philippians," *Biblica* 60, no. 1 (1979): 99.

preposition and prefix, as that is the focus of our argument regarding the emphasis they place on how believers connect to Christ and to one another. The combination of the *syn* preposition and prefix in compound form appears from beginning to end in Paul's letter. Using the structural analysis of the letter as set out by Fee, we have *syn*-prepositions and prefixes in every major section of the epistle.[13]

Introductory Matters (1:1-11)
syn (1:1)
synkoinōnos (1:7)

Paul's Affairs: Reflections on Imprisonment (1:12-26)
synechō (1:23a)
syn Christō (1:23b)

The Philippians' Affairs: Exhortation to Steadfastness and Unity (1:27–2:18)
synathleō (1:27)
sympsychos (2:2)
synchairō (2:17, 18)

What's Next: Regarding Paul's and Their Affairs (2:19-30)
syn (2:22)
synergos (2:25a)
systratiōtēs (2:25b)

Their Affairs Again (3:1–4:3)
symmorphizō (3:10b)
symmimētēs (3:17)
symmorphos (3:21)
syzygos (4:3a)
syllambanō (4:3b)
synathleō (4:3c)
synergos (4:2d)

Concluding Matters (4:4-23)
synkoinōneō (4:14)
syn (4:21)[14]

---

[13]Fee, *Philippians*, 54-55.
[14]We think this structural analysis, with the *syn*-compounds embedded in each major section of the epistle, meets Macaskill's context and narrative requirement of Paul's use of the prefix *syn* relative to our union with Christ. See Macaskill, *Union*, chapters 9–11.

HORIZONTAL *SYN*-COMPOUNDS IN PHILIPPIANS 1

Paul signals the significance of the horizontal *syn*-compound in the epistle when he uses one in the middle of the opening thanksgiving as the descriptor of his relationship to this group of believers: *synkoinōnous mou tēs charitos*, meaning "partners together in grace" (Phil 1:7, our translation). This very same *syn*-compound (*synkoinōnēsantes*) appears in the letter's conclusion, where Paul thanks the church because they were "partners together in my affliction" (Phil 4:14, our translation). Between these *syn*-compound bookends, Paul repeatedly uses other *syn*-compounds to focus his attention on the need for this church to remain connected to Christ and to one another in the face of internal dissention and external opposition. Paul addressed the reason for this emphasis on the connections among church members in his letter to the Ephesians, where he demonstrates in Ephesians 4:16 that horizontal connections between believers convey spiritual nutrients necessary for their continued growth. In Philippians, the horizontal *syn*-compounds serve as supporting pillars to Paul's overall concern for the church "to stand firm for the gospel and to be united in Christian love."[15]

In the unique *syn*-compound (*syn* + *koinōnia*) in Philippians 1:7, we have two of the most important terms in the Pauline motif of connectedness evident in the epistle. Although scholars differ as to the precise translation of the phrase here in Philippians 1:7, we can say with some confidence that Paul opens the letter with a note of gratitude for their "participation (*koinōnia*) in the gospel" in his current imprisonment and in his general "defense and confirmation of the gospel" (Phil 1:5, 7). In regard to his ministry circumstances, Paul is grateful that the Philippians "are partakers of grace with me" (Phil 1:7, 12). Paul views the Philippians as *synkoinōnoi*— "partners together" or "participants together" in the grace of God manifested in all the circumstances of his apostolic ministry. Paul uses a similar phrase with this *syn*-compound in 1 Corinthians 9:23, where he describes himself as a "partner with the gospel" (*synkoinōnos*, our translation). Morna Hooker diverges from traditional interpretations of this phrase because of the "force of the *syn*" that leads the translation to be not "participant in the

---

[15]Peter T. O'Brien, *Epistle to the Philippians: A Commentary on the Greek Text* (Grand Rapids: Eerdmans, 1991), 38.

gospel" but "fellow participant in the gospel," thereby emphasizing how "*others* may share in the benefits *with me*."[16] Unless prevented by the context, the force of the *syn*-compound here and elsewhere in Philippians emphasizing "with me" and "others" should carry the same weight as Hooker gives it in 1 Corinthians 9:23.

The Philippians have manifested consistent support of Paul, regardless of his circumstances, and this support has manifested itself in the *koinōnia* gift mentioned in his closing thanks ("shared with me," Phil 4:15). The *koinōnia* language Paul uses in Philippians 1:5 and 7 and in Philippians 4 lends support for understanding the basis of his appreciation for their partnership with him in his ministry. They exhibited their partnership with Paul by supporting him financially during his imprisonment.[17] At that time, as in many poorer countries today, relatives and friends must provide food, clothing, and other resources needed to sustain the prisoner through incarceration.[18] Such assistance the Philippians shared with Paul would supplement the meager rations provided by the Roman state.

The *synkoinōnos* language Paul uses in Philippians 1:7 (a cognate noun) and Philippians 4:14 (a cognate verb) to describe the spiritual connection between Paul and the Philippians reveals important aspects of the mutual spiritual benefits such a partnership provides. The mutual participation of Paul and the Philippians in God's grace "prompted the Philippians to alleviate Paul in his imprisonment" and sustained Paul during his confinement.[19] It certainly seems clear from Philippians 1:7 that Paul also embraces, within the scope of this spiritual connection he has with the Philippians through Christ, their participation not only in his imprisonment but also in the "defense and confirmation of the gospel." Paul does not elaborate on exactly what role the Philippians played in this, but we know he thought they were in some sense involved with him in his apostolic ministry. Such a general

---

[16]Morna D. Hooker, "A Partner in the Gospel: Paul's Understanding of His Ministry," in *Theology and Ethics in Paul and His Interpreters: Essays in Honor of Victor Paul Furnish*, ed. Eugene H. Lovering Jr. and Jerry L. Sumney (Nashville: Abingdon, 1996), 85-86.

[17]Julien M. Ogereau, "A Survey of Koinonia and Its Cognates in Documentary Sources," *Novum Testamentum* 57 (2015): 275-94.

[18]Norval Morris and David J. Rothman, *The Oxford History of the Prison: The Practice of Punishment in Western Society* (Oxford: Oxford University Press, 1995), 19-21.

[19]O'Brien, *Epistle*, 70.

description of the connection between Paul and the Philippians leaves the application to contemporary ministry settings much more fluid and flexible.

When we collaborate with (*synkoinōnos*) others in any ministry effort (i.e., teaching, preaching, witnessing, serving, leading, or worshiping) as connected members of the body of Christ, there is mutual spiritual benefit conveyed between the partners. This seems to be the point Paul is trying to convey to the Philippians when he uses this same *syn*-compound (*synkoinōneō*) as he closes his letter in Philippians 4:14. Following his description of the Philippians as those who "partner together with me in my affliction," he describes the *koinōnia* (what the Philippians shared with Paul) as a "giving and receiving" exchange (Phil 4:15). Although the Philippians gave the gift for Paul's benefit, he is interested in "the profit which increases to your account" (Phil 4:17). Scholars have noted the use of commercial bookkeeping language here and elsewhere in Philippians. However, as Peter Marshall informs us, the phrase "in the matter of giving and receiving" (*eis logon doseōs kai lēmpheōs*) was a familiar idiom expressing friendship. Marshall explains that in this context "it reflects a warm and lasting relationship. He not only receives the gift gladly as a sign of their continuing concern, but also recalls the mutual exchange of services and affection which they had shared in the past."[20] Our connections to one another as fellow Christians derive from our mutual connection to Christ as head of the body, creating spiritual conduits of mutual exchange and mutual spiritual benefit. Paul uses *syn*-compound and *koinōnia* language as part of a motif of connectedness to express these "mutual exchanges of services and affection" to which Marshall alluded.

Fee notes that the force of the *syn* + *koinōnia* in Philippians 1:7 translates as "to participate in something," and in this context it means "to be participants together in the grace of God."[21] Expanding upon Pauline usage in general, Dunn suggests, "Paul's language indicates rather a quite profound sense of participation with others in a great cosmic movement of God centered on Christ and effected through his Spirit."[22] These two specific

---

[20]Peter Marshall, *Enmity in Corinth: Social Conventions in Paul's Relations with the Corinthians* (Tübingen: Mohr Siebeck Verlag, 1987), 163-64.
[21]Fee, *Philippians*, 91n87.
[22]Dunn, *Theology*, 404.

interpretations fit nicely with J. Y. Campbell's more general understanding of *koinōnia* as "participation in something in which others also participate" with the "something" being supplied by the context in which *koinōnia* occurs.[23] They also complement Julien Ogereau's exhaustive analysis of *koinōnia* and its cognates applied to Paul's letter to the Philippians.[24] In particular, concerning the *synkoinōnos* language Paul uses in Philippians 1:7 and 4:14, Ogereau offers considerable evidence to suggest that the idea of a "partnership" between Paul and the Philippians dominates the message of the epistle. The word conveys the idea of "their cooperative and supportive role as his close associates in his evangelistic endeavors."[25] He argues that the *koinōnia* cognates understood as partnership "constitute *the* interpretive key that unlocks a proper understanding of Paul's relationship with the Philippians."[26] This lends support to our argument that Paul uses a combination of *koinōnia* cognates and *syn*-compounds to form a motif of connectedness between himself and the Philippians as well as among the Philippians themselves.

We should not overlook the fact that this description of Paul's relationship with the Philippians with a horizontal *syn*-compound (*synkoinōnia*) sets up Paul's "Intercession for the Philippians' Spiritual Growth" in the prayer that follows (Phil 1:9-11).[27] Nor should we ignore the appreciation Paul expresses to this congregation in Philippians 1:19 for their prayers on his behalf, which he expects will result in his "deliverance" and "the provision of the Spirit of Jesus Christ." Although the syntax is tricky, the most attractive interpretation understands the phrase "through your prayers" (*dia . . . deēseōs*) as the means (*dia* as a preposition of agency) God will use to deliver Paul and give him a "generous supply" (Fee's translation, *epichorēgia*) of the Spirit.[28]

One of the ways in which spiritual partners connect to one another in a spiritual ecology and share spiritual resources is by praying for one another. When we intercede on behalf of a brother or sister to God in prayer, we are

---

[23]J. Y. Campbell, "*Koinonia* and Its Cognates in the New Testament," *Journal of Biblical Literature* 51 (1932): 352-82, 353.

[24]Julien M. Ogereau, *Paul's Koinonia with the Philippians: A Socio-Historical Investigation of a Pauline Economic Partnership* (Tübingen: Mohr Siebeck, 2014).

[25]Ibid., 311.

[26]Ibid., 311.

[27]Frank Thielman, *Philippians*, NIV Application Commentary (Grand Rapids: Zondervan, 1995), 40.

[28]Fee *Philippians*, 130-34, and O'Brien, *Epistle*, 110-12.

engaging in intercessory prayer. This is what Frank Thielman means by his term *intercession.*[29] Paul has a *syn*-compound for almost every spiritual relationship that can exist between believers. He has one for Christians praying together too. In 2 Corinthians 1:8-11, Paul describes the way in which God delivered him during so many of his ministry exploits when he and his coworkers "despaired even of life" (2 Cor 1:8). During these threatening moments of ministry, Paul knew he was not alone because the Corinthians were "laboring together (*synypourgountōn*) through prayer" (2 Cor 1:11, our translation). At the end of the epistle to the Romans, Paul entreats them "to strive together with me (*synagōnisasthai*) in your prayers to God for me" (Rom 15:30). Praying together and praying for one another is in Paul's mind one of the most important ways Christians utilize the spiritual connections they have with one another in Christ. The language he used to describe the activity of praying with and for suggests that in his mind it is a strenuous and vigorous effort. The *synagōnizō* compound combines *syn* with *agōn*, a Greek word that comprises part of a larger "*agōn* motif" in Paul's epistles.[30] It is a motif or cluster of synonyms all built around the athletic imagery of ancient Greece and Rome. All of the terms that comprise the motif in Paul's letter "suggest the thought of exertion and maximum endeavor."[31] Prayer for Paul was not a passive folding of the hands in a serene posture of worship but an active exertion of maximum effort in collaboration with other Christians. He viewed it as a legitimate engagement with him in ministry and often requested that his coworkers in ministry pray specifically for him as he prayed for them (2 Cor 13:9; Eph 1:16; Phil 1:4; Col 1:3, 9; 1 Thess 1:2; 5:25; 2 Thess 3:1-2). Paul's prayer relationship with his churches and coworkers was a reciprocal exchange of ministry effort that produced mutual spiritual benefit for all involved. Paul prayed for his churches and coworkers and he asked them to pray for him. Gordon Wiles, in his definitive treatment of Paul's intercessory prayers, offers evidence "that there had existed a consciously maintained network of mutual intercessions between Paul and his churches."[32] Wiles describes a spiritual network involving spiritual interconnections

---

[29]Thielman, *Philippians*, 40.

[30]Victor C. Pfitzner, *Paul and the* Agon *Motif* (Leiden: E. J. Brill, 1967).

[31]Ibid., 10.

[32]Gordon P. Wiles, *Paul's Intercessory Prayers: The Significance of the Intercessory Prayer Passages in the Letters of Paul* (Cambridge: Cambridge University Press 1974), 278.

between those who are "co-participants in the gospel" and "partners together in God's grace" (Phil 1:5, 7, our translation). As Wiles observes in his introduction to his study of Paul's intercessory prayers, Paul "wanted his churches to grow rapidly toward maturity . . . His intercessory prayers seem to be closely related to a longing for the maturity of his churches, as the scope of his own mission extended."[33]

The central role of intercessory and reciprocal prayer in Paul's mission alerts us to its strategic importance for the process of spiritual formation within the spiritual ecology of the body of Christ. Through our spiritual connections to each other, we convey spiritual nutrients as we pray with and for one another (Jas 5:16).

In Philippians 1:27 and 4:3 we find another set of *syn*-compounds (*synathleō*) that open and close the epistle. Although the terms are identical, Paul uses them with different referents in mind. In Philippians 1:27 he expresses his hope that whether he is absent or present, "I will hear of you [plural] that you [plural] are standing firm in one spirit, with one mind striving together (*synathleō*) for the faith of the gospel." The *syn*-compound here and the word *agōn* in Philippians 1:30 ("the same conflict which you saw in me") signals to Victor Pfitzner that the athletic imagery suggesting a struggle should not be inflated to the level of armed combat or "a head-on frontal battle between the faith, and its enemies."[34] Instead, the language suggests the common struggle Paul and the Philippians share in "the spread and growth of faith."[35] He feels confirmed in this interpretation by the manner in which Paul uses the same *syn*-compound in Philippians 4:3 coupled with the *syn*-compound *synergoi* ("workers together"), a term Paul typically used in reference to those who partner with him in gospel ministry of various kinds.

HORIZONTAL *SYN*-COMPOUNDS IN PHILIPPIANS 2

In Philippians 2, we find three horizontal *syn*-compounds that capture our attention. It is hard to say a great deal about *sympsychoi* "together in spirit" because it is the only occurrence of this *syn*-compound in the New Testament

---

[33]Ibid., 3.
[34]Pfitzner, *Paul and the* Agon *Motif*, 116.
[35]Ibid., 117.

(Phil 2:2, our translation).[36] The Philippians can manifest the "fellowship (*koinōnia*) of the Spirit" that they already possess and make "[Paul's] joy complete" by "being of the same mind, maintaining the same love, united in spirit, intent on one purpose" (Phil 2:2). The adjective *sympsychos* describes how the Philippians are to reveal such unity. The compound form of the word expresses the idea of "together as one person" that puts the emphasis on their internal connectedness as a group of believers.[37] Robert Jewett explores the word's usage outside the New Testament and suggests that it expresses how the Philippians share a common existence. He proposes that the word possesses a strong social dimension expressing group solidarity in which the Philippian congregation is "joined in one common life."[38] If the Philippians are to withstand the external opposition and the internal squabbles and make Paul's joy complete, they must be predisposed to a mindset (*phroneō*) of themselves as a cohesive, spiritually connected and unified community (*sympsychoi*).

Christians connected to Christ (vertical *syn*-compounds) and to one another (horizontal *syn*-compounds) have spiritual ties that create a vast web of interconnections that, through the Spirit, form them into a *sympsychoi*— a unified soul or person. Americans and others from the Global West have difficulty conceiving of individuals forming a single whole persona. Thinking ecologically helps us grasp this concept embedded in Paul's use of this phrase and others for a collective social unit. An ecosystem consists of parts that comprise a whole. All of the individual believers in the church at Philippi are to think of themselves as one person or one soul, rather than a collection of individuals who just happen to believe the same thing and worship at the same church. Paul thinks ecologically about the church and this perception manifests itself in the language he uses to describe it.

HORIZONTAL *SYN*-COMPOUNDS IN PHILIPPIANS 3

In Philippians 3:17, Paul calls on the Philippians to "join in following my example" (*symmimētēs*) and uses a *syn*-compound that encourages the

---

[36]Although this is the only occurrence of *syn* + *psychē*, there are other occurrences of the *psychē* root in the epistle (Phil 1:27; 2:19, 20, 30).

[37]Fee, *Philippians*, 185.

[38]Robert Jewett, *Paul's Anthropological Terms: A Study of Their Use in Conflict Settings* (Leiden: E. J. Brill, 1971), 350.

church as a whole to "be imitators together" or "co-imitators" of Paul. Here we see the only occurrence of this *syn*-compound in Paul's epistles and in the whole of Greek literature, although he uses *mimētēs* in noncompound form (1 Cor 4:16; 11:1; Eph 5:1; 1 Thess 1:6; 2:14; 2 Thess 3:7, 9).[39] Since this is the only occurrence of this particular *syn*-compound, we must consider its meaning and significance within the setting of the entire epistle in which Paul uses a significant number of *syn*-compounds to "emphasize his and their mutuality."[40] The consensus among those who have analyzed this word in its context is that "Paul is calling on them to join together in imitating him."[41] Imitating Paul was a way for the Philippians to express their connective unity in Christ by performing a collective action.[42] The appeal Paul makes for the Philippians to be co-imitators of him is "yet another way to urge the community to be united in their life in Christ."[43] The collective action suggested by *symmimētēs* "implies the notion of transfer of character or personality from one person to another."[44] In this case, since the action involves the entire Philippian community, we might modify Sanders's description of imitation to involve the transfer from one person to other people who are, collectively, to imitate him as the exemplar is imitating Christ (1 Cor 11:1; 1 Thess 1:6). The force of the two previous "let us" imperatives in Philippians 3:15-16 suggests that Paul is thinking of the entire church at Philippi on the same terms as his coworkers (*synergoi*), who together imitate him and the "pattern" (3:17, our translation, *typon*) Paul has provided for them to replicate.

The specifics of what the Philippians together are to imitate in Paul is influenced by the preceding context. Paul recounts his personal attitude about how he "count[s] all things to be loss in view of the surpassing value of knowing Christ Jesus my Lord" and how he presses on "toward the goal for the prize of the upward call of God in Christ Jesus" (Phil 3:8, 14). This pattern he asks them to imitate together as one is not just of himself because

---

[39]Fee, *Philippians*, 363.

[40]Ibid., 363n10.

[41]Ibid., 365; see also Hooker, "Partner."

[42]Adele Reinhartz, "On the Meaning of the Pauline Exhortation '*mimētai mou ginesthe*'—Become Imitators of Me," *Studies in Religion* 16, no. 4 (1987): 393-403; O'Brien, *Epistle*.

[43]Hansen, *Letter*, 261.

[44]Boykin Sanders, "Imitating Paul: 1 Cor 4:16," *Harvard Theological Review* 54 (1981): 358.

he encourages them to "observe those who walk according to the pattern you have in us" (Phil 3:17). Thus we have a collective model (Paul and the others) combined with a collective admonition ("let us") directed toward a collective imitation (*symmimētēs*). The corporate aspects of the appeal dominate this and other portions of Paul's letter to the Philippians, undergirded by his use of horizontal *syn*-compounds such as we have here in Philippians 3:17.

HORIZONTAL *SYN*-COMPOUNDS IN PHILIPPIANS 4

In Philippians 4:3, Paul finishes off his flurry of horizontal *syn*-compounds with a blizzard of four in one verse. We want to tease out the significance of each one, but before we do, we should set the context. Beginning in Philippians 4:1, Paul finishes off his letter with many of the themes or issues addressed earlier. His general admonition to the church to "stand firm in the Lord" is counterbalanced by a specific admonition directed to Euodia and Syntyche to "live in harmony in the Lord" (Phil 4:1, 2). In the same way, Paul's previous general admonition to the church in Philippians 2:2-4, to be "of the same mind, maintaining the same love, united in spirit, intent on one purpose . . . with humility of mind regard one another (*allēlōn*) as more important than yourselves . . . [while looking out for] the interests of others," is now specifically directed to two women.

As Paul targets the two named women, which is unusual for Paul, he unleashes a cascade of horizontal *syn*-compounds that form the basis of his specific appeal to unity in general and in the case of Euodia and Syntyche. We have highlighted the four *syn*-compounds in the following translation of Philippians 4:3: "Indeed, true companion (*syzyge*), I ask you also to help (*syllambanou*) these women who have shared my struggle (*synēthlēsan*) in *the cause of* the gospel, together with Clement also and the rest of my fellow workers (*synergōn*), whose names are in the book of life."

Following Fee, we take "true companion" as one in a series of four *syn*-compounds that Paul uses to make his point for seeking unity generally and specifically as it applies to the two women he names.[45] Paul uses the term (*syzygos*) to refer to a role a person will assume in relation to the two women.

---

[45]Fee, *Philippians*, 392n40.

The word in compound form is used extensively outside of the New Testament to refer to those who are "yoked together" or are "partners" in some relationship (business, politics, marriage), or what we might refer to as a *colleague*.[46] The term *"zygos"* referred originally to "a bar or frame of wood by which two draft animals are joined at the head or neck in order to work together effectively in pulling a plow, harrow, or wagon."[47] The intensified form with a *syn* prefix emphasizes the connectedness of two joined together in some common effort. Paul might be thinking of this coworker as the bar or frame of wood who acts as "conciliator" between the two women.[48] Paul calls upon this person to serve as a go-between and bring about reconciliation. We can only speculate as to what caused the disruption in their personal relationship, but whatever it was it apparently spilled over into the community and had the potential of great harm to the church's unity and cohesion.

In addition to acting as an agent of reconciliation (2 Cor 5:18-21), the unidentified coworker of Paul is to "help these women" (4:3), identifying our second horizontal *syn*-compound in the cluster (*syllambanou*). The specific task involved here indicated by *syllambanou* is that of assisting the two women as they work through their differences to reach a satisfactory solution.[49] Balz and Schneider suggest the translation "grasp together," which certainly fits the context where Paul is asking this coworker to serve as a connecting yoke between two disconnected women and bring them back together into a reconciled relationship.[50] This translation is not far off from one of three main uses of the sixteen occurrences of the *syn*-compound in the New Testament where the word occurs in context of physical seizing or grasping (Mk 14:48; Lk 5:7; Acts 12:3). While the physical aspect may be missing from Philippians 4:3, it suggests toward these two women the action Paul directs the coworker to initiate. Paul wants this unnamed person to

---

[46]Joseph Henry Thayer, *Greek-English Lexicon of the New Testament* (New York: American Book Company, 1889), 594.

[47]Johannes P. Louw and Eugene A. Nida, *Greek-English Lexicon of the New Testament: Based on Semantic Domains*, 2nd ed. (New York: United Bible Societies, 1989), 54.

[48]Fee, *Philippians*, 392n40.

[49]D. H. Field, "Conceive, Apprehend," in *The New International Dictionary of New Testament Theology*, ed. Colin Brown (Grand Rapids: Zondervan, 1979), 1:344.

[50]Horst Balz and Gerhard Schneider, eds., *"Syllambánō,"* in *Exegetical Dictionary of the New Testament* (Grand Rapids: Eerdmans, 1994), 285.

conciliate a reconciliation by bringing two separated women back together and restoring the disconnection between them.

Paul indicates the motivation or rationale for this action of restoration by his use of the third *syn*-compound, *synathleō*, meaning "struggle together" or "compete together." The word as it appears here and in Philippians 1:27 is a compound of *syn* + *athleō*, which Victor Pfitzner includes in Paul's "*agōn* motif" where he uses athletic and military imagery to express "the thought of exertion and maximum endeavor" he and others expended for the cause of the gospel.[51] He uses this *syn*-compound in Philippians 1:27 in reference to the common struggle of the entire congregation (probably at its founding) and here in Philippians 4:3 with direct reference to the two women, Clement, and the other unnamed coworkers. The term suggests "an active participation in Paul's own wrestling for the spread of the faith of the Gospel."[52] All of the athletic imagery that appears in Philippians and elsewhere (e.g., 1 Cor 9:24-27) conveys the importance of a singular (Phil 3:13, *hèn dé*) goal-orientation (Phil 3:14, *skopon*).[53] Today we often hear people admonish others to "keep your eyes on the prize," and that is exactly what Paul encourages the Philippians to do. Both spiritual formation and mission have Christ as their focus (Phil 1:18-29; 3:7-16). Concerning Paul's mission, the goal is that "Christ will be exalted" and regarding Paul's concept of the goal of spiritual formation "that I may gain Christ" and be "conformed to His death" (Phil 3:8, 10). As Bradley Arnold says, "Since the goal for Paul is Christ, exploring what he emphasizes about Christ in Philippians can help us to understand the ways in which this goal might shape how he runs."[54] It also shapes how he encourages others to run and compete in the contest of gospel proclamation, witness, and transformation. If all involved remain disconnected and not unified in mind (*phroneō*, "one mindset") and heart, it is impossible to reach a common goal. Paul's frequent use of *phroneō* for "mindset" or "attitude" (Phil 1:7; 2:2, 5; 3:15, 19; 4:2, 10) "is to enable the community to live in unity and harmony, despite the pressures from within and without."[55]

---

[51]Pfitzner, *Paul and the* Agon *Motif*, 10.

[52]Ibid., 119.

[53]Bradley Arnold, "Re-Envisioning the Olympic Games: Paul's Use of Athletic Imagery in Philippians," *Theology* 115, no. 4 (2012): 243-52.

[54]Ibid., 248.

[55]David G. Horrell, *Solidarity and Difference: A Contemporary Reading of Paul's Ethics* (London: T&T Clark, 2005), 214.

The entire phrase with *synathleō* is "who have shared my struggle in the gospel." Paul packs a great deal into this single phrase. First, he elevates the role of these two women in his mission by identifying them along with men named in this epistle (Clement, Timothy, Ephaphroditus) and in Romans 16 as his coworkers in the gospel. What specific ministry they performed Paul does not tell us, but we do know from this *syn*-compound that it was a significant role that involved ministry partnership with Paul. Hansen notes, "These two women were full members of his mission team, *along with Clement and the rest of my co-workers*."[56] Second, because Paul writes this way, we have to conclude that Paul is thinking of more than just the effect on the church; he is also considering the impact on his own gospel mission (*en tō euangeliō*). If these two women, who have a ministry connection with others like Clement in Paul's mission, remain in such a broken relationship, it may spread throughout the ranks of Paul's considerable host of coworkers and hamper the forward movement (3:14, *diōkō*) he envisioned for himself and the mission (Phil 3:5-16).

We find our final horizontal *syn*-compound in Philippians at 4:14: "Nevertheless, you have done well to share *with me* (*synkoinōnēsantes*) in my affliction." Fee notices the friendship language Paul uses here and in other places in Philippians by reference to the "giving and receiving" mentioned in Philippians 4:15, indicative of the presupposition of reciprocity that undergirds the chief mark of friendship.[57] Such friends or partners in ministry and mission not only share materials with one another (see Phil 4:17-18) but share or partner together in affliction (*thlipsis*). Fee notes the force and significance of the *syn*-prefix that although "redundant . . . in particular emphasizes their participating together with him *in his affliction*."[58] He characterizes the *syn* as "redundant" because Paul affixes it to *koinōneō*, which in its own right conveys the idea of sharing or partnership. The *syn*-prefix heightens or intensifies the verb *koinōneō* so that the Philippians do not miss Paul's message about how strongly he feels about their support during his "affliction," which is probably a reference to his imprisonment (Phil 4:14). The significance of this for Paul is likely due to the fact that the Philippians

---

[56]Hansen, *Letter*, 285.
[57]Fee, *Philippians*.
[58]Ibid., 438n8.

shared together in his affliction during the time of their own affliction (Phil 1:28-30). This is exactly the mindset he recommended to them through the example of Christ, his own experiences, and those of Timothy and Epaphroditus (Phil 2:5-11 NIV; see also Phil 1:12-20; 2:19-30).

The particular form the joint participation in Paul's affliction assumed at the hand of the Philippians appears in Philippians 4:17-18, where Paul describes it as a "fragrant aroma" and "acceptable sacrifice" that is "well-pleasing to God." Here Paul moves from the imagery of business and finance applied to friendship to that of sacrifice. The language Paul uses originates in the sacrificial texts of the Old Testament and expresses Paul's attitude toward what the Philippians did by sharing with him and God's attitude toward it as suggested in the phrase "well-pleasing to God." The shift to sacrificial imagery may reflect the larger New Testament vision of God's people as a "holy priesthood" and "royal priesthood" that now offers "spiritual sacrifices acceptable to God through Jesus Christ" (1 Pet 2:4-10 NIV). Hebrews carries forward this view, enjoining the Christian community to "continually offer up a sacrifice of praise to God . . . and do not neglect doing good and sharing (*koinōnia*), for with such sacrifices God is pleased" (Heb 13:15-16). Paul views the Philippians as his partners in mission and ministry and also as part of a new priesthood enacted in the work of Christ who satisfied God's requirements and brought to an end the need for animal sacrifices (Heb 10:1-25). Those joined to Christ as "a high priest forever according to the order of Melchizedek" are qualified to offer spiritual sacrifices to God and to one another (Heb 6:20; 13:15-16; Phil 4:17-18).

When we "share together" or "partner together" (*synkoinōnos*), as the Philippians did with Paul in his affliction, by sharing (*koinōneō*) spiritual sacrifices, we cultivate our ecological connections to one another and contribute reciprocally to each other's spiritual formation. Paul engaged in intercessory prayer for the Philippians, and they "partnered together in the gospel" and "partnered in grace" with him (Phil 1:3-7, our translation). The Philippians obeyed Paul, and Paul poured himself out "as a drink offering upon the sacrifice and service" of their faith (Phil 2:12, 17). Paul sent them Timothy and Epaphroditus, and the Philippians sent him a sacrificial gift (Phil 2:19-30; 4:15-18). The mutual participation of connected members of the body of Christ produced an ample supply; Paul and the Philippians received from Paul the promise that God would "supply all your needs according to His riches in glory in Christ Jesus" (Phil 4:18-19). This mutual benefit occurs when God's people cultivate the spiritual connections that join us together with Christ and one another.

Paul's concern that the Philippian church stay connected to Christ and to each other through the Spirit mirrors a similar concern he expresses about the church in Corinth. Although, as Fee points out, the situation in the Philippian church had not yet reached critical mass as it did in Corinth, Paul is concerned to avoid that catastrophe and "head them off at the pass."[59] Why are the social and interpersonal relationships between members of his congregations such a major concern? Paul recognized that if social and interpersonal disconnections remained and festered into open fragmentation (*schismata*, 1 Cor 1:10-11), each congregation risked jeopardizing their own spiritual welfare and ability to build one another up toward the Pauline goal of transformation (*metamorphoō*, 2 Cor 3:18).

James Thompson refers to the Pauline goal as "Paul's pastoral ambition . . . to participate with God in effecting the transformation of his communities."[60] He proposes that we consider this "pastoral theology of transformation . . . as the center of Paul's thought."[61] He demonstrates that Paul's use of "the language of formation" built upon the frequency of words formed from the Greek root *morph* "indicates the central role of this concept in Pauline theology."[62] Thompson examines several Pauline epistles to establish his thesis, but we are interested only in his analysis of Paul's use of this *morph*-based language in Philippians. In this regard, he notes that "Paul's use of forms of *morph*- indicates the significance of transformation in this letter."[63] Paul uses *morphē* in Philippians 2:6 and 7 to refer to the "form of God" Jesus possessed in his eternal existence and the "form of a bond-servant" Jesus assumed in his incarnation (Phil 2:6-7). Thompson

---

[59]Ibid., 385-98.

[60]James W. Thompson, *Pastoral Ministry According to Paul: A Biblical Vision* (Grand Rapids: Baker Academic, 2006), 20-23.

[61]Ibid., 20.

[62]The Greek root *morph* ("form, shape") forms the basis for *morphē* ("visible form," Phil 2:6, 7), *morphoō* ("to shape or form," Gal 4:19), *morphōsis* ("outward form," 2 Tim 3:5), *symmorphos* ("have the same form," Phil 3:21), *symmorphizō* ("be conformed with," Phil 3:10), *metamorphōsis* ("transform," Mark 9:2; Rom 12:2; 2 Cor 3:18). Ibid., 23.

[63]Ibid., 47.

concludes, "As the repetition of forms of *morph-* indicates, the language of transformation provides the structure of the passage."[64] Indeed, Thompson says, "Philippians offers a comprehensive look at Paul's understanding of transformation as the basis for his pastoral work."[65] One of the important insights Thompson gleans from his study of transformation in Philippians (and in other Pauline epistles) is the challenge it offers to our contemporary, individualistic, and privatized concept of spiritual formation. He sees that, in Philippians, Paul prefers a "corporate formation" that embraces the individual as a connected member of a larger body.[66] He suggests that "transformation occurs when the community is shaped by the sacrifice of the cross" as the church community seeks to replicate, through the power of the Spirit, "the mind of Christ."[67]

SIGNIFICANCE FOR A SPIRITUALLY NETWORKED ECOLOGY

We need not leave Paul's rich insight about the power of the interconnections we have with one another through Christ to theologians and ministers. If we only think about the significance of what he is teaching us in theological or ministry categories, we deprive ourselves and others of its benefit across all sectors of our human ecologies. In particular, if we want to see the lives of our online students spiritually enriched and ultimately transformed, we must take Paul's description of our relationships with one another in Christian community seriously. In practice, this means we must attend to the critical ecological elements of interconnections and interactions leading to mutual growth. K. Patricia Cross argues that all learning involves making connections of various sorts. Specifically, she identifies cognitive, social, and experiential connections.[68] The latter is especially important when working with adults who always view learning through the lens of their own experiences. Online courses need opportunities for students to connect with other students and faculty, to connect God's Word to course concepts, to connect old knowledge to new information, and to connect action with reflection.

---

[64]Ibid., 47.
[65]Ibid., 53.
[66]Ibid., 154.
[67]Ibid., 151.
[68]K. Patricia Cross, *Learning Is About Making Connections*, The Cross Papers #3 (Mission Viejo, CA: League for Innovation in the Community College, 1999).

In addition to learning connections, we also need a variety of learning interactions. Michael Moore proposed three types of online interactions: learner-content, learner-instructor, and learner-learner.[69] Building on Moore's taxonomy of interaction we might add instructor-instructor, instructor-content, learner-group, group-group, instructor-group, group-content, and learner-LMS (learning management system) interaction. Course designers need to assess online courses with these two ecological axes in view. We believe that courses with high levels of a creative variety of connections and interactions will more likely enhance and promote whole-person development than those that ignore or limit these ecological elements.

Paul's *syn* vocabulary reveals his deep insight about how Christians connect to Christ in all aspects of his work of salvation and also about how Christians connect to one another to produce a rich harvest of spiritual benefits for all. One might even say, based upon his horizontal *syn* vocabulary, Paul thinks of each local church as a *collaboratory* of relationships, worship, and ministry where God's people visibly live out their identities as God's spiritually networked ecology. The task of all connected believers within this spiritual ecology is to find collaborators who complement one another in giftedness and mission orientation. The post-Enlightenment concept of ordained leaders who guide lay members to perform ministry functions as part of a programmatic structure geared to numerical growth cuts against the grain of what Paul envisions. Although he held apostolic authority, he rarely exercised that authority unless pushed to do so by the unruly behavior of specific congregations, as in Corinth. Paul's preference was to work collaboratively and flexibly with partners who shared his vision of a harmoniously functioning ecology of interconnected and gifted members. Käsemann argues that regardless of whether authority finds its grounding in charisma or office, the final test of its authenticity is "that which edifies the community."[70] Paul trusted the Spirit to work through these organic connections to empower, encourage, and edify individually and collectively. He practiced what Jesus taught his disciples about how leaders lead most effectively when they serve others rather than demanding

---

[69]Michael G. Moore, "Three Types of Interaction," *American Journal of Distance Education* 3, no. 2 (1989): 1-4.

[70]Ernst Käsemann, *Essays on New Testament Themes* (Philadelphia: Fortress, 1964), 67.

that others serve them. Like Jesus, Paul turned human concepts of leadership on their head, placing the leader at the service of those situated within their sphere of spiritual influence. When we understand the ecology of the body of Christ, as Paul described it, we gain a deeper appreciation of the power of ministry partnerships rooted in our common partnership with Christ. We strengthen and edify one another through these ministry connections, and, ultimately, we glorify Christ by living out through our relationships the truth of who we are in him. As Brian Hedges put it so precisely, "Spiritual renewal doesn't happen in isolation from others. It happens in relationships."[71]

In addition to these vital ecological connections, Paul envisions a highly interactive and transactive dynamism that functions between and among these connections. Here we learn to appreciate Paul's use of the "one another" (*allēlōn*) language as his way of conceptualizing what transpires between connected believers who share a common connection to Christ. The New Testament uses the reciprocal pronoun *allēlōn* to describe the variety of ways we interact with one another through the connections represented by the horizontal *syn*-compounds. We now turn to a study of these "one another" interactions to learn about their role in stimulating spiritual growth through the spiritual ecology of the church.

---

[71]Brian G. Hedges, *Christ Formed in You: The Power of the Gospel for Personal Change* (Wapwallopen, PA: Shepherd Press, 2010), 238.

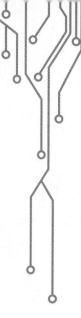

ELEVEN

ECOLOGICAL INTERACTIONS WITH OTHER CHRISTIANS

Relationships are the ecology of God's kingdom, the new creation.

LEONARD SWEET, *OUT OF THE QUESTION . . . INTO THE MYSTERY*

Do you remember playing on a teeter-totter on the playground as a youngster? Maybe you were like us and tried to get the beam to balance perfectly level for as long as possible. They always were lots of fun especially when both kids were willing to work together cooperatively. However, when one kid decided to jump off and leave the other to crash to the ground, suddenly all the fun was gone! Relationships with other Christians work like those playground teeter-totters. They have a certain give-and-take, up-and-down quality to them, require equilibrium, and only work if both parties are cooperative. In this chapter, we want to examine the critical role that our relationships to other Christians play in the spiritual formation of the body of Christ.

The horizontal *syn*-compounds the apostle Paul used in his epistles describe the spiritual connections Christians have with each other by our shared connection to Christ as head of the church. The spiritual ecology of the church manifests the same connections and interactions that God's created ecologies possess. The organic connections that link different living organisms in a natural ecology provide a potential for mutual growth through shared nutrients and resources. The connections require activation and utilization, however, if *potential* growth is to become *actual* growth.

The connections serve as nutrient pathways, but something must pass between the connections for all of the connected organisms to benefit. The same process of ecological growth is present in the spiritual ecology of the Christian community. The spiritual connections we have with other Christians in Christ (see previous chapter) require activation and utilization through an interactive process of shared spiritual nutrients and resources provided by God's Word through God's Spirit. In this chapter, we will highlight how spiritually connected Christians can engage each other in mutual and spiritually beneficial ways.

RECIPROCAL INTERACTIONS IN NATURAL ECOSYSTEMS

The study of ecosystems in nature has demonstrated, both at a conceptual and trophic level, that there are ongoing cybernetic exchanges of energy, materials, and nutrients that benefit all interconnected entities.[1] The study of the "exchange of energy and materials between organisms and the environment" is commonly known as "ecosystem ecology," a subdiscipline of ecology.[2] This area of ecological specialization studies the exchange of nutrients, such as water, carbon, nitrogen, and other critical elements for growth.

R. Margalef defined ecology as "the study of systems at a level in which individuals or whole organisms may be considered elements in interaction, either among themselves, or with a loosely organized environmental matrix."[3] He then described the cybernetic feedback loops produced from "elements linked by reciprocal influences."[4] The ecosystem as a whole consists of many interacting parts, which "therefore reciprocally influence and determine each other."[5]

---

[1] Bernard C. Patten, "Network Ecology: Indirect Determination of the Life-Environment Relationship in Ecosystems," in *Theoretical Studies of Ecosystems: The Network Perspective*, ed. M. Higashi and T. P. Burns (New York: Cambridge University Press, 1991); A. R. E. Sinclair and Charles J. Krebs, "Trophic Interactions, Community Organization, and Kluane Ecosystem," in *Ecosystem Dynamics of the Boreal Forest: The Kluane Project*, ed. Charles J. Krebs, Stan Boutin, and Rudy Boustra (New York: Oxford University Press, 2001).

[2] Yiqi Luo, Ensheng Weng, and Yuanhe Yang, "Ecosystem Ecology," in *Encyclopedia of Theoretical Ecology*, ed. Alan Hastings and Louis J. Gross (Berkeley: University of California Press, 2012), 219.

[3] R. Margalef, "Networks in Ecology," in *Theoretical Studies of Ecosystems: The Network Perspective*, ed. M. Higashi and T. P. Burns (New York: Cambridge University Press, 1991), 4.

[4] Ibid., 7.

[5] Jan Christian Smuts, *Holism and Evolution* (Gouldsboro, ME: The Gestalt Journal Press, 1986), 86.

The result of these various reciprocal exchanges and interactions between individual entities in an ecosystem is the general viability and sustained growth and development of both the entities and the ecosystem. Even negative influences transmitted through the interactions often strengthen defense mechanisms as plants learn to adapt to the presence of initially inhibitory substances.[6]

RECIPROCAL INTERACTIONS IN HUMAN ECOSYSTEMS

One of the leading bioecological developmentalists over the last forty years, Urie Bronfenbrenner, proposed his ecology of human development model as a way to explain the interactive process of human development across time.[7] He was critical of the traditional paradigm for studying human development that viewed subjects like test-tubes in a laboratory instead of observing them in real-life settings. Instead, he offered a wider ecological view that took place in social contexts, such as the family, church, school, and neighborhood. He likened these interconnected contexts to a "set of nested structures, each inside the next, like a set of Russian dolls."[8] He envisioned human development as a social ecology in which each individual human connected to other human beings through social ties that required interactive exchanges of various types.

At the heart of Bronfenbrenner's understanding of human growth was the developmental motor of reciprocity. He defined *reciprocity* as a pattern of ongoing "concomitant mutual feedback" that over time "motivates the participants not only to persevere but to engage in progressively more complex patterns of interaction, as in a ping-pong game in which the exchanges tend to become more rapid and intricate as the game proceeds."[9] Although recognized as a necessary component of human development, reciprocity, according to Bronfenbrenner "is often disregarded in practice."[10] Ironically, we will find a similar disregard for this necessary component of spiritual development.

---

[6]S. J. H. Rizvi and V. Rizvi, eds., *Allelopathy: Basic and Applied Aspect* (London: Chapman and Hall, 1992).

[7]Urie Bronfenbrenner, *The Ecology of Human Development: Experiments by Nature and Design* (Cambridge, MA: Harvard University Press, 1979).

[8]Ibid., 3.

[9]Ibid., 57.

[10]Urie Bronfenbrenner, "Toward an Experimental Ecology of Human Development," *American Psychologist* (July 1997): 519.

According to Bronfenbrenner, reciprocal interactions between developing persons in various sectors of the social ecology (family, church, school) "are especially salient both as influences on and manifestations of development."[11] In fact, he argues that the developmental potential of a person in a defined social ecology is enhanced when properly motivated "to engage in progressively more complex . . . patterns of reciprocal interaction . . . with others in that setting."[12]

The net result of all of these various types of reciprocal interactions is "reciprocal development" among all people who are emotionally bonded, which introduces "further complexity into the developmental equation."[13] Reciprocal activity, reciprocal interactions, and all types of reciprocal exchanges between people in a variety of social ecological environments benefit all of the interconnected and interacting members developmentally. This is the "further complexity" to which Bronfenbrenner refers that recognizes the power such social interactions and exchanges have on the development of participants. The developmental influence is bidirectional and even multidirectional when one considers the different ecological layers in Bronfenbrenner's Russian doll illustration. These developmental realities create a web of reciprocal interactions across all of the interconnected social ecologies that influence to one degree or another the mutual development of all connected and interacting persons. Coupled with the emotional bond created through the reciprocal interactions, we have a powerful social ecosystem that nurtures human development to its fullest God-given potential.

RECIPROCAL INTERACTIONS IN BONHOEFFER

Although Dietrich Bonhoeffer never met or read Bronfenbrenner, his doctoral dissertation, published under the title *Sanctorum Communio*, has much in common with Bronfenbrenner's concept of reciprocal interactions and interconnections in a social setting.[14] As Clifford J. Green notes,

---

[11]Bronfenbrenner, *Ecology*, 55.
[12]Ibid., 163.
[13]Ibid., 65-66.
[14]Dietrich Bonhoeffer, *Sanctorum Communio: A Theological Study of the Sociology of the Church* (Minneapolis: Fortress, 1998). Originally published in Germany in 1930, republished in 1954, 1960, and 1969 then published in English in 1963 from the third German edition. A critical edition that included material excised in the first edition (1930) appeared in Germany in 1986. The text we are working from is an English translation of this critical text that appears as volume one of *Dietrich Bonhoeffer Works*.

Bonhoeffer "articulates the concept of . . . the reciprocal relationship of person and community."[15] Bonhoeffer's dissertation was an attempt to integrate the study of sociology and theology to form a comprehensive theology of the Christian community. As Bonhoeffer wrote in one of the sections of his original dissertation, which ended up on the editing room floor, "The essence and structure of the new humanity 'in Christ' can only be understood in contrast to humanity 'in Adam'; one must know of the general and specific social relations of humanity in order to understand the empirical phenomenon of the church sociologically."[16] In spite of the scholarly dissertation quality to his writing at this early stage of his career, we can discern Bonhoeffer's methodology. He draws from the then-current study of sociology to clarify his biblical and theological insights about the Christian community and how its members grow together in Christ. Although Bonhoeffer provides a very dense and reasoned explanation of his concept of the Christian community, we limit our study to his remarks about the relationships between Christian people and the Godhead that create our distinctive community. In particular, he focused on two terms from the sociological literature that helped explain what he observed in Scripture. The first term, *Wechselwirkungen*, referred to the social "interactions" between persons, and the second term, *Wechselbeziehungen*, referred to the "reciprocal relations" between human beings.[17] He understood from these two terms that to some extent "human beings are determined . . . by living in interaction with other people."[18]

Human community is possible as long as there is a "mutual harmony of reciprocally directed wills," but what happened in the fall was a disruption of this harmony with egocentric self-seeking.[19] This initial disruption produced social fragmentation and the *peccatorum communio*, or "community of sinners," in Adam.[20] This historic disconnection from God and other human beings produced a form of community, but it was one that could never achieve the original divine purpose. We humans lost our capacity for "mutual harmony of reciprocally directed wills," and in so doing, "each of us has turned to his own way" (Is 53:6; cf. Is 47:15; 56:11).

---

[15]Ibid., 1.
[16]Ibid., 201n5.
[17]Ibid., 25-26.
[18]Ibid., 26.
[19]Ibid., 117.
[20]Ibid., 118.

It is "through Christ" that Christians have the capacity to "have access to one another, joy in one another, and fellowship with one another."[21] As a result of Christ's work to create a new, redeemed community, the Holy Spirit makes the many into one, thereby maintaining the unity of the church as a community of faith. Faith in Christ makes possible entry into this community, and "the Holy Spirit brings us into community with God."[22] The unique ministry of the Holy Spirit—nothing else—produces unity in community "through co-operative activity and reciprocal influence."[23] Part of the cooperative activity involves "mutual service" in which each member of the community "becomes in practice a priest for the other."[24] This sacrificial service for the other produces "mutual affection" among the members who love one another.[25] This mutual service and mutual love results in the building up of the church "with a growth which is from God" (Col 2:19).

None of Bonhoeffer's thoughts about Christian community were purely theological or theoretical. We know that in the seminary community in Finkenwalde he practiced what he preached to others. In October 1938, shortly after the arrest of Martin Niemoller, Bonhoeffer addressed a group of young theologians, urging them to remain true to Christ and his word in the face of persecution and suffering. In part, he reminded them of the need for "reciprocal action" toward one another, the importance of their "care for one another," and the necessity to "suffer with another" as members of the body of Christ.[26] He closed this portion of his remarks by reminding his audience that Paul wrote to Timothy from prison "not to be ashamed of his fetters but to suffer with him" and noted that this was "the last appeal we have from the Apostle Paul."[27]

RECIPROCAL INTERACTIONS IN OLD TESTAMENT COVENANTS

The importance of reciprocal relationships can also be found in the Old Testament. Daniel Elazar argues that "in all its forms, the key focus of covenant

---

[21]Ibid., 39.
[22]Ibid., 165.
[23]Ibid., 194.
[24]Ibid., 207.
[25]Ibid., 246.
[26]Dietrich Bonhoeffer, *The Way to Freedom: Letters, Lectures and Notes 1935-1939* (New York: Harper & Row, 1966), 182.
[27]Ibid., 183.

is on relationships."[28] Every biblical covenant creates a covenant obligation "designed to reinforce mutuality" between the covenant partners.[29] Jacob Milgrom gets more specific when he notes that "the reciprocal relationship between God and Israel is one of the hallmarks of the covenant" summarized in Deuteronomy 26:17-18: "You have today declared the LORD to be your God. . . . The LORD has today declared you to be His people, a treasured possession."[30] Nelson Glueck wrote the definitive study on the Hebrew covenant word *ḥesed* and offers valuable insights that enhance our understanding of reciprocity in biblical covenants.[31]

Glueck defines *ḥesed* as "mutuality or reciprocal conduct" or "a relationship of mutual reciprocity."[32] He placed primary emphasis in the word upon conduct and behavior in a relationship rather than attitude (mercy) or feeling (love), which he regarded as "peripheral aspects of *ḥesed*."[33] The relationship may be based in kinship, friendship, hospitality, or citizenship. The two key words in his concept of *ḥesed* are "mutuality" and "reciprocity," treated as synonyms by Glueck. Being in a covenant relationship, whatever form it takes, always implied mutual obligations for the parties involved. This is why Glueck insists that "*ḥesed* was the content of every *bᵉrith*."[34] Building upon Glueck's concept of *ḥesed*, Lester Kuyper offered a description more familiar to the language of the New Testament when he described it as "fellowship established between two or more people."[35] Those in fellowship through a covenantal bond exhibited behavior consistent with *ḥesed* and expected equivalent behavior in return. Thus, *ḥesed* expressed various types of reciprocal interactions between persons that were manifestations of their mutual obligations to each other.

Both Glueck and Kuyper use the story of Ruth, Naomi, and Boaz from the book of Ruth to illustrate how *ḥesed* manifested itself in obligated

---

[28]Daniel J. Elazar, "The Political Theory of Covenant: Biblical Origins and Modern Developments," *Publius* 10, no. 4 (1980): 6.

[29]Ibid., 27.

[30]Jacob Milgrom, "Profane Slaughter and a Formulaic Key to the Composition of Deuteronomy," *Hebrew Union College Annual* 47 (1976): 10.

[31]Nelson Glueck, *Hesed in the Bible* (Jersey City, NJ: KTAV, 1975).

[32]Ibid., 39, 71.

[33]Ibid., 60.

[34]Ibid., 73.

[35]Lester J. Kuyper, "Grace and Truth," *Reformed Review* 16, no. 1 (1962): 3.

relationships. Ḥesed appears as a major theme in the book as Ruth shows ḥesed to her deceased husband and to her mother-in-law, Naomi, while Boaz shows ḥesed to Ruth and Naomi as well as Ruth's deceased husband. Throughout the book of Ruth, we see evidence of a reciprocal relationship that proved to be mutually beneficial to both women. The reciprocal and mutual nature of the relationship between Ruth and Naomi even surfaces in the literary structure of the book.[36] There is equal space given to both women, with Naomi as the focus of attention in the opening and closing chapters of the book and Ruth as the focus of attention in the two middle chapters. This literary structure gives prominence to both women equally. Opening and closing with Naomi indicates her literary importance, but placing Ruth in the middle of the structure highlights her as equally essential to the purposes of the author. This is evidence of an intentional balancing act to give each woman her place without sacrificing the other.

We find in the story a powerful reciprocal relationship between two women brought together by family tragedy. In their common widowhood, Naomi advises her daughter-in-law to return home like her sister, Orpah, because Naomi thinks it would be in Ruth's best interest to do so. Ruth responds with her iconic pledge to Naomi and the God of Israel.[37] Ruth goes into the field to glean in order to provide food, and Naomi gives Ruth advice on how to approach their nearer kinsman, Boaz. Ruth gives Naomi a son, but Naomi nurses him. The reciprocal nature of their relationship reaches its climax in Ruth 4 at the birth of Obed, the son born of the union of Ruth and Boaz. It seems as though the author means to show Obed as the son of *both* Ruth and Naomi. In Ruth 4:13 Obed is clearly identified as Ruth's son, but in Ruth 4:17 the women say, "A son has been born to Naomi!" The net effect is that the story teaches that both women were ultimately redeemed and enjoyed the mutual benefits not only of their reciprocal relationship to each other but also their mutual relationship to Boaz the kinsman-redeemer.

---

[36] Athalya Brenner, "Naomi and Ruth," *Vetus Testamentum* 33, no. 4 (1983): 385-97.

[37] "Intreat me not to leave thee, or to return from following after thee: for whither thou goest, I will go; and where thou lodgest, I will lodge: thy people shall be my people, and thy God my God" (Ruth 1:16 KJV).

RECIPROCAL INTERACTIONS IN THE BODY OF CHRIST

We have seen that reciprocal interactions play a vital role in facilitating growth in natural ecosystems from a number of sources: in the growth of human beings in social ecologies through the work of Bronfenbrenner, in Bonhoeffer's theology of community, and in covenantal relationships, such as that of Ruth, Naomi, and Boaz, where covenant partners seek each other's best interests. Reciprocal relationships were an integral part of the social glue that, along with the bonded connection to Christ through the Spirit, held the diverse New Testament church together. The covenantal demands of *ḥesed*, incumbent upon all parties to a covenant, obligated them to reciprocal behaviors that demonstrated their mutual commitment. As recipients of a new covenant relationship in Christ, New Testament believers come under similar reciprocal behavioral expectations. In Hebrews 10:11-18, we find a direct reference to the New Covenant of Jeremiah 31. Beginning at Hebrews 10:19, we find a series of admonitions ("Therefore, brethren") that flow from this common covenantal connection—one of which is "let us consider how to stimulate one another to love and good deeds" (Heb 10:24). The Greek word translated "one another" is *allēlōn*, a reciprocal pronoun that conveys mutual obligation.

Paul, along with other New Testament writers, employs the reciprocal pronoun *allēlōn* to describe how connected believers relate to each other as members of the new covenant community formed through the blood of Christ (Lk 22:20).[38] The reciprocal pronoun *allēlōn* occurs exactly one hundred times in the New Testament, spanning every collection of New Testament writing and every author. Rarely are just two people involved in a "one another" relationship (Lk 23:12; 24:14, 17, 32; Acts 15:39; 1 Cor 7:5). Most often, the reciprocal pronoun refers to "groups of persons who are in some way peers and with reference to relationships within a homogeneous group in order to express communication with or, sometimes, negative conduct toward each other."[39] H. Krämer's description accurately conveys

---

[38]Although *heautou* ("each other") is a synonym of *allēlōn* ("one another"), for sake of space we delimit our analysis to the occurrences of *allēlōn*, although there is much profit in including it in a more thorough treatment of the terms.

[39]H. Krämer, "*Allēlōn*," in *Exegetical Dictionary of the New Testament*, ed. Horst Balz and Gerhard Schneider (Grand Rapids: Eerdmans, 1999), 1:63.

what we prefer to call an ecological use of the term that recognizes the interconnective and interactive nature of relationships between connected members of the body of Christ. Although, as Krämer indicates, some of the occurrences of the term appear in contexts that express "negative conduct toward each other," we limit our analysis to the positive uses of the word.[40]

THE MEANING OF *ALLĒLŌN* IN THE NEW TESTAMENT

Lexical definitions are a helpful starting point for determining the meaning of a word, but they are not the most decisive. Thayer defines *allēlōn* as "one another, reciprocally, mutually," while Louw and Nida define it as "a reciprocal reference between entities—'each other,' 'one another.'"[41] A. T. Robertson comments that "this pronoun brings out the *mutual* relations involved."[42] In the New Testament, the term *allēlōn* conveys a relationship between two or more people committed to one another through a common faith in Jesus Christ. The relationship is interactive, with each person contributing to the welfare of the other through a variety of behaviors and attitudes. For instance, Christ commanded his followers to "love one another" (Jn 13:34). This particular *allēlōn* imperative serves as the umbrella description for all subsequent *allēlōn* commands. Christians express their love for one another as they "build up one another," "pray for one another," "[forgive] one another," "greet one another," "serve one another," "comfort one another," "stimulate one another to love and good deeds," and "[admonish] one another" (2 Cor 13:12; Gal 5:13; Eph 4:32; Col 3:16; 1 Thess 4:18; 5:11; Heb 10:24; Jas 5:16). As Fong observes, "This pronoun brings out the mutualness of interrelations. The persons involved in the given context respond to each other in like fashion. They share in the same activity toward one another."[43]

---

[40]Ibid., 63.

[41]Joseph Henry Thayer, *Greek-English Lexicon of the New Testament* (New York: American Book Company, 1889), 28; Johannes P. Louw and Eugene A. Nida, *Greek-English Lexicon of the New Testament: Based on Semantic Domains*, 2nd ed. (New York: United Bible Societies, 1989), 816.

[42]A. T. Robertson, *A Grammar of the Greek New Testament in the Light of Historical Research* (Nashville: Broadman Press, 1934), 692.

[43]Bruce William Fong, "Our Mutual Responsibilities Toward One Another as Seen in the Context of *Allēlōn*" (unpublished master of theology thesis, Dallas Theological Seminary, May 1978), 8.

RECIPROCAL INTERACTIONS

In spite of the numerous "one another" admonitions in the New Testament, and especially in Paul's epistles, most commentators have largely ignored the significance of this pronoun. Gerhard Lohfink notes, for example, that "Kittel and Friedrich's ten-volume *Theological Dictionary of the New Testament,* which treats even individual prepositions, does not consider *allēlōn* worth an entry."[44] One exception to this oversight is Gordon Fee, who writes in regard to Paul's concept of the people of God, "This concern for God's saving a people for his glory is further demonstrated by the frequency of one of the most common, but frequently overlooked words in Paul's ethical exhortations: *allēlōn*. Everything is done *allēlōn*."[45] Fee argues that the Holy Spirit works through these reciprocal interactions among members of the body "creating a people for his name, among whom God can dwell and who in their life together will reproduce God's life and character in all its unity and diversity."[46]

Each of the reciprocal imperatives that use *allēlōn* has a verb that describes the attitude or behavior one Christian is to exhibit to another. When one eliminates duplicate verbs from the hundred occurrences of *allēlōn* in the New Testament, there are twenty-nine distinct verbs that identify ways Christians reciprocally interact with each other. Some of these verbal expressions, such as "build up one another" or "encourage one another" (1 Thess 5:11), are more general, while others, such as "greet one another with a holy kiss" (Rom 16:16) or "be hospitable to one another" (1 Pet 4:9), are more specific. The wide range of attitudes and behaviors enumerated in the different verbs reflects the multitude of ways Christians can engage one another in relationships, ministry, worship, social encounters, prayer, and various manifestations of Christian fellowship.

RECIPROCAL COMMUNITY

One of the consequences of the various reciprocal interactions discussed above is the formation of a reciprocal community of Christians who are

---

[44]Gerhard Lohfink, *Jesus and Community* (Philadelphia: Fortress, 1984), 99.
[45]Gordon Fee, *God's Empowering Presence: The Holy Spirit in the Letters of Paul* (Peabody, MA: Hendrickson, 1994), 871.
[46]Ibid., 872.

spiritually connected to Christ and one another. We borrow the term "reciprocal community" from Jack Balswick et al., who refer to the reciprocal community as "the true church—as the body of Christ."[47] In their understanding of human and spiritual development, the authors identify reciprocity as "the glue that holds the relational polarities . . . together."[48] The polarities of individual uniqueness and group cohesion and unity can only exist together when each individual lives in "reciprocating, authentic relationships with others."[49] The capacity to do this, they argue, derives from the *imago Dei*, in which "bearing the image of God means living as unique individuals in reciprocating relationships with others."[50] Likewise, Stanley Grenz situates the capacity for community in "the Christian understanding of God," in that "God's triune nature means that God is social or relational— God is the 'social Trinity.'"[51] He then demonstrates that "the social nature of the divine image" manifests itself among his creatures as "a special enjoyment of community . . . as we live in fellowship" with other similarly endowed humans.[52]

Lohfink demonstrates that Jesus seeks to fulfill the failed mission of Israel to bring God's salvation to the ends of the earth through a faithful remnant of twelve disciples and others.[53] The mission of God must have a people, called by his name, who declare his glory among the nations, and in this respect Jesus forms a community of disciples who will carry forward the mission of Jesus after his earthly mission is complete. It is this "little flock" to whom Jesus entrusts his mission in the Great "Co-Mission" of Matthew 28:16-20 (see Lk 12:32). Lohfink grounds the formation of this new Jesus community in the *allēlōn* language first used by Jesus and then picked up by Paul and others.[54] When Jesus instructed his remnant community to "wash one another's feet" and to "love one another, even as I have loved you," he

---

[47]Jack O. Balswick, Pamela Ebstyne King, and Kevin S. Reimer, *The Reciprocating Self: Human Development in Theological Perspective* (Downers Grove, IL: IVP Academic, 2005), 53.

[48]Ibid., 35.

[49]Ibid., 36.

[50]Ibid., 36.

[51]Stanley J. Grenz, *Created for Community: Connecting Christian Belief with Christian Living*, 2nd ed. (Grand Rapids: Baker Books, 2001), 52.

[52]Ibid., 79.

[53]Lohfink, *Community*.

[54]Ibid., 99-106.

established the baseline imperative for the manifestation of an already existing community (Jn 13:14, 34). It remained for Paul and others to explicate precisely how Christ's followers were to love one another through the multitude of *allēlōn* imperatives sprinkled throughout the rest of the New Testament. Lohfink argues that Paul is interested in carrying forward the mission of Jesus by founding "edifying communities" in which "Paul thought of *edification* primarily in connection with the local community."[55] Paul's way of expressing how this edification would be achieved was, in part, reflected in his use of the *allēlōn* imperative originally directed to the formation of a new community comprised of Jews and Gentiles (Eph 4:2, 25, 32; 5:21).[56]

RECIPROCAL SPIRITUAL GROWTH

While necessary for community formation and beneficial relationships, reciprocal interactions have a teleological orientation in that they should ultimately produce reciprocal spiritual growth. This is why Lohfink places his discussion of *alllēlōn* within his section on edification.[57] Reciprocal relationships that strengthen reciprocal communities should ultimately effect reciprocal growth among all participating members. What is true of organic growth in natural ecologies and human growth in social ecologies also seems true of Christian growth in spiritual ecologies.

Whereas the *syn*-compounds reveal how *connected* we are to Christ and to one another; the *allēlōn* imperatives reveal how *contagious* we are to one another. Through reciprocal interactions with other Christians, we mutually influence one another toward maturity and a sharper reflection of the image of Christ, whether in person or through online formats. We think this is the tenor of Proverbs 27:17 (NIV): "As iron sharpens iron, so one person sharpens another." The "one another" interactions we have with other Christians are spiritually sharpening, producing mutual spiritual benefit. The "one another" interactions we have with other learners are intellectually sharpening. Although the reciprocal interactions may be one-on-one encounters, as envisioned in Proverbs, they may also be one-to-many, as with Paul and

---

[55]Ibid., 100.
[56]Carl B. Hoch, "The Significance of the *Syn*-Compounds for Jew-Gentile Relationships in the Body of Christ," *Journal of the Evangelical Theological Society* 25, no. 2 (1982): 175-83.
[57]Lohfink, *Community*, 99-106.

the church at Rome, or many-to-many, as in Paul's comments to the church in Thessalonica (Rom 1:11-12; 2 Thess 1:3). Whatever form they take (and wherever they occur), the mutual spiritual interactions produce mutual spiritual influence. The interactions are spiritual in the sense that they are empowered by the Spirit and they help influence spiritual and learning outcomes (love, peace, harmony, unity, etc.).

We see the connection between reciprocal interactions and reciprocal growth in Romans 1:11-12. Paul opened his letter to the church at Rome by expressing his desire to see them so "that I may be encouraged together with you (*symparaklēthēnai*), each of us by one another's (*allēlōn*) faith, both your faith and my faith" (Rom 1:12, our translation). Marty Reid argues that this passage "introduces Paul's rhetoric of mutuality, which encompasses the various facets of his purpose in writing."[58] Paul is attempting to establish the foundation of his relationship to the church in Rome before he arrives in person. Although he is "called to be an apostle" (Rom 1:1 NIV), he wants them to understand that they are connected to one another through Christ— hence his use of the *syn*-compound (*synparakaleō*)—and that they have the potential to mutually influence one another spiritually—hence the use of the *allēlōn* pronoun in Rom 1:12. Yes, Paul will share his faith with them, but they will also share their faith with him so that they will be encouraged together ("both your faith and my faith"). Just as Paul opened the epistle with the *allēlōn* pronoun, he closes with a flourish of ten *allēlōn* imperatives from Romans 12–16, including "*be* devoted to one another," "love one another," edify one another, "admonish one another," and, finally, "greet one another with a holy kiss" (Rom 12:10; 13:8; 14:19; 15:14; 16:16). Through the performance of these "one another" behaviors, those connected to Christ "stimulate one another" to a more mature faith (Heb 10:24).

In chapter eight we introduced you to the concept of whole-person transformation into the fullness of Christ as the targeted outcome of the spiritual formation process. In that chapter, we provided the hand illustration to depict the various aspects that comprise the whole person. We identified six aspects—physical, intellectual, emotional, social, moral, and spiritual— that are indicative, but not necessarily exhaustive, of what constitutes a whole

---

[58]Marty L. Reid, "A Consideration of the Function of Romans 1:8-15 in Light of Greco-Roman Rhetoric," *Journal of the Evangelical Theological Society* 38, no. 2 (1995): 190.

person. If we listed all of the *allēlōn* exhortations and then superimposed our hand of human wholeness over them, we would find each of the six dimensions represented (see fig. 4).

Accept one another
Bear one another's burdens
Forbear one another
Be at peace with one another
Live in harmony with one another
Be hospitable to one another

Be kind to one another
Be devoted to one another
Encourage one another
Show concern for one another
Comfort one another

Do not lie to one another
Do not slander one another
Do not judge one another
Honor one another

Moral
Social
Emotional
Intellectual
Spiritual

Teach one another
Speak to one another
Stimulate one another
Admonish one another

Confess your sins to one another
Pray for one another
Build up one another
Forgive one another
Love one another
Submit to one another

Physical

Wash one another's feet
Greet one another with a holy kiss
Wait for one another

Figure 4. Paul's "one another" commands and the six developmental dimensions

This suggests to us that the practice of the reciprocal interactions within a community of Christ-followers has the potential of producing reciprocal formation among the interacting people. The fact that they encompass the breadth of what the Scriptures seems to teach regarding Paul's description of "the fullness of Christ" and how Christians together are "to grow up in all *aspects* into Him" indicates that they are necessary for achieving the targeted outcome of the spiritual formation process (Eph 4:13, 15).

Reciprocal interactions with others produce reciprocal growth. Our various ways of engaging and experiencing each other in different settings throughout our lives facilitate or instigate the process of human development.[59] The same patterns of growth that researchers observe in human growth finds a parallel in patterns described in the New Testament. In the

---

[59]Bronfenbrenner, *Ecology*.

same way that we need reciprocal interaction with others to experience reciprocal human development, we also need reciprocal interaction with other Christians to experience reciprocal Christian development.[60] God designed the reciprocal nature of the imperatives to provide opportunities for Christians to interact, engage, and exchange actions and attitudes in a symbiotic give-and-take empowered by the Holy Spirit. We do not engage in these activities so we can simply enjoy one another's fellowship, although this is an obvious outgrowth of the interaction. We do not relate to one another reciprocally so we can simply worship with integrity and enjoy authentic Christian relationships, although this is an important aspect of their use. Everything—fellowship, relationships, and worship—leads to and contributes to our whole-person transformation into the fullness of Christ. The ultimate outcome of reciprocal relationships and interactions between believers is individual and corporate transformation into the image of Christ. The process that produces this ultimate outcome is reciprocal development that unfolds over a lifetime of faithful obedience. For this reason, Howard concludes from his own analysis of the role of the believing community in spiritual maturity that "in a very real way, believers need one another."[61] The *allēlōn* imperatives express how this need is met and the mutual edification that results.

INTENTIONAL CONNECTIONS THROUGH ONLINE EDUCATION

Online learning environments offer a unique opportunity for the kind of reciprocal interactions Paul describes, Bronfenbrenner studied, and Bonhoeffer practiced. In fact, we have found from our experience that these kinds of interactions are more intentional and deliberately planned in online courses than in traditional classroom settings. A study of natural, social, human, and spiritual ecology reminds us that nothing grows and matures in isolation. Growth only occurs through organic or social connections supplemented by reciprocal interactions that spread needed nutrients among the connected members of the ecosystem. As Hunter writes, based upon the hermeneutical pattern established in the parables of Jesus, "What

---

[60]Balswick, King, and Reimer, *Reciprocating Self*, 285-97.

[61]James M. Howard, *Paul, the Community, and Progressive Sanctification: An Exploration into Community-Based Transformation within Pauline Theology* (New York: Peter Lang, 2007), 187.

is valid in one sphere is valid also in the other."[62] We have seen that the connections and interactions required for natural and human growth pertain also to spiritual growth in the ecology of the body of Christ. As students find themselves connecting and interacting over course content and often over non-course-related encouragement and support, they experience a "one another" lifestyle that will not only benefit their personal growth but positively affect their immediate family, local church, and community. In the next chapter, we will see that these reciprocal interactions, reflected in the *allēlōn* imperatives scattered throughout the New Testament, create an environment in which we spread the contagion of holiness, one Christian to another, through the empowerment of the Holy Spirit.

---

[62] Archibald M. Hunter, *Interpreting the Parables* (Philadelphia: The Westminster Press, 1960), 8.

— TWELVE —

ECOLOGICAL SANCTIFICATION

We might thus speak of holiness for Jesus . . .
being that which he views as "contagious."

CRAIG L. BLOMBERG, *CONTAGIOUS HOLINESS*

Since we live in a highly interconnected universe, we also live in a highly contagious universe. Ebola outbreaks in Africa and the SARS virus in Asia serve as constant reminders that diseases spread quickly in our highly interconnected world. The movie *Contagion* (2011) dramatically depicted the panic that ensues when pandemics threaten thousands and possibly millions of lives. Cures can spread just as quickly, as we learned after Jonas Salk discovered the polio vaccine in 1955 and virtually eliminated this dreaded childhood disease. The contagion effect, however, is not limited to biological viruses. As we saw in chapter eight, social contagions can spread through a social network, infecting all connected people with ideas, beliefs, values, behaviors, and norms. Biological and social contagions serve as paradigms for understanding spiritual contagion that spreads through connected members of the Jesus network. Just as trees influence one another in forest ecologies through natural connections, so do human beings influence one another in social ecologies through social connections. Researching ecosystems, B. C. Patten discovered "the power of networks to influence organisms" and concluded from his study that "the key property of the network phenomenon is *influence*."[1] The influence we have on each other through our

---

[1] B. C. Patten, "Network Ecology: Indirect Determination of the Life-Environment Relationship in Ecosystems," in *Theoretical Studies of Ecosystems: The Network Perspective*, ed. M. Higashi and T. P. Burns (Cambridge: Cambridge University Press, 1991), 292.

social networks spreads from one person to another as a social contagion.[2] Likewise, the spiritual influence we exert on our brothers and sisters in Christ through our spiritual connection to Christ (see chapter nine) and our spiritual connections to one another (see chapter ten) spreads through the ecology of the body of Christ like a spiritual contagion.

CONTAGIOUS HOLINESS

An example of this kind of spiritual contagion is the contagious nature of holiness Jesus modeled in his public ministry. Jesus revolutionized Jewish understandings of what it meant to be holy as God's people. N. T. Wright explains this when he writes, "In some Jewish purity codes, uncleanness was regarded as infectious; Jesus' actions (touching a leper, for example) challenged this, indicating that it was genuine holiness, not uncleanness that was infectious."[3] According to Wright, Jesus "redefine[d] the concept of holiness."[4] The prevailing view of established Judaism in the first century was that holiness is something to quarantine from contamination. Religious leaders, such as the Pharisees, thought holiness was fragile and susceptible to contamination and easily overpowered by uncleanness and impurity. The best way to protect it, they argued, was to hedge it in and avoid any contact with potential sources of contamination, such as sinners, Gentiles, tax collectors, and Jews who did not observe purity laws that protected God's holiness.

Some, like the Essenes, even physically separated themselves from other observant Jews who, in their view, were not observant enough. They established the Qumran community in the Judean wilderness and thereby separated themselves from the defilement of the Jewish temple and the city precincts of Jerusalem.[5] Among the Dead Sea Scrolls, the Temple Scroll

---

[2]Everett M. Rogers, *Diffusion of Innovations*, 5th ed. (New York: Free Press, 2003); Albert-László Barabási, *Linked: How Everything Is Connected to Everything Else and What It Means for Business, Science, and Everyday Life* (London: Plume, 2003); Jonah Berger, *Contagion: Why Things Catch On* (New York: Simon & Schuster, 2013); Nicholas A. Christakis and James H. Fowler, *Connected: The Surprising Power of Our Social Networks and How They Shape Our Lives* (New York: Little, Brown, 2009).

[3]N. T. Wright, foreword to *Conflict Holiness and Politics in the Teachings of Jesus*, rev. ed., by Marcus J. Borg (Harrisburg, PA: Trinity International Press, 1998), xv.

[4]Ibid., xxiii.

[5]James VanderKam and Peter Flint, *The Meaning of the Dead Sea Scrolls: Their Significance for Understanding the Bible, Judaism, Jesus, and Christianity* (New York: HarperOne, 2002); Robert Eisenman, *The Dead Sea Scrolls and the First Christians* (Shaftesbury, UK: Element Books Limited, 1996).

envisioned a new temple built by "the sons of light," where they could preserve and protect God's holiness from "the sons of darkness."[6]

Amid these dominant attitudes that governed the thinking of Pharisees,
Levites, and Sadducees, Jesus introduced a radical inversion. God's holiness
was not fragile and easily contaminated; it was powerful and contagious.
Holiness had the intrinsic power to overcome cultic impurity and human
uncleanness and make people clean and pure. Jesus' attitude toward those
bound by impurity and uncleanness was not a judgmental attitude but one
of mercy and compassion. He understood that they were in the throes of a
power that deformed them in different ways—physical, emotional, psychological, and spiritual. He offered them deliverance through a holiness that
transformed them completely in all of those dimensions.

The focal point of establishment Judaism and its deep concern about God's
holiness was the meal table. As Borg reports, "No fewer than 229 of the 341
rabbinic texts attributed to the Pharisaic schools of Shammai and Hillel
pertain to table fellowship."[7] They were most concerned about matters of ritual
purity (washing hands), tithing, and the observance of kosher food preparation and service. These religious leaders built a ritual fence around their
tables that kept out any potential source of defilement. Included in this table
barrier were those serving the food and guests eating the food. Traditional
Jewish leaders would limit participation at their dining table to those like
themselves who strictly observed these holiness codes regarding table fellowship. Thus in Luke 5, when Jesus accepts Matthew's invitation to attend "a
big reception for Him in his house" and at this meal were "a great crowd of tax
collectors and other *people* who were reclining *at the table* with them," we are
not surprised when the Pharisees and scribes begin to grumble (Lk 5:29-30,
emphasis added). The crucial question they posed was, "Why do you eat
and drink with the tax collectors and sinners?" (Lk 5:30). Tax collectors
and sinners were people who disregarded ritual purity observances and
were thus unclean and contaminated. By "reclining at the table" with these
violators of God's holiness, those who considered themselves holy were

---

[6]Hershel Shanks, *Understanding the Dead Sea Scrolls* (New York: Random House, 1992), 126-29.
[7]Marcus J. Borg, *Conflict, Holiness, and Politics in the Teachings of Jesus*, rev. ed. (Harrisburg, PA:
  Trinity International Press, 1998), 95.

thereby contaminated and defiled.[8] From the perspective of the scribes and Pharisees, the contagious uncleanness and ritual defilement of the tax collectors and sinners infected and ritually contaminated Jesus and his disciples.

Craig Blomberg surveys the major New Testament texts where Jesus sits at the table with social and religious outcasts:

> The unifying theme that emerges from the passages surveyed . . . is one that may be called "contagious holiness." Jesus regularly associates with various sorts of sinners on whom the most pious in his culture frowned . . . he does not assume that he will be defiled by associating with corrupt people. Rather, his purity can rub off on them and change them for the better.[9]

Hanna Stettler confirms Blomberg's analysis by concluding that Jesus "sanctified people by having table fellowship with them, thereby bringing them back into communion with God."[10]

Jesus scandalized his critics even further in Luke 5 when he says, "*It is* not those who are well who need a physician, but those who are sick" (Lk 5:31, emphasis added). The Great Physician comes into physical contact with the contaminated sick, but instead of catching their disease, he infects them with his holiness and they are healed. We see this portrayed in the encounter Jesus has with the leper, the dead body of Jairus's daughter, and the dead body of the widow of Nain (Mk 1:41; 5:41; Lk 7:14). This is exactly what we see happening with the woman who has the blood issue; she too is ritually impure and unclean, and yet she touches the hem of Jesus' garment (Mk 5:25-34). Amazingly, instead of her touch contaminating Jesus, a power goes out from him and heals her. His holiness is more contagious than her uncleanness. Hanna Stettler concludes from these accounts of direct contact with Jesus that "his holiness must be transferable."[11]

The most remarkable event that illustrates the contagious holiness of Jesus is Mark 5 and Jesus' encounter with the so-called Gerasene Demoniac—the wild man he met in the tombs on "the other side of the sea" (Mk 5:1). In

---

[8]Reclining at table was a Greco-Roman practice adopted by Jews as the typical social and physical arrangement for dining. A triclinium was a three-sided table arranged among other similarly designed tables where participants would recline on pillows and eat in a leisurely manner.

[9]Craig L. Blomberg, *Contagious Holiness: Jesus' Meals with Sinners* (Downers Grove, IL: InterVarsity Press, 2005), 128.

[10]Hanna Stettler, "Sanctification in the Jesus Tradition," *Biblica* 85 (2004): 162.

[11]Ibid., 160.

the first century, the phrase "the other side of the sea" referred to the Gentile section known as the Decapolis (Mk 4:35; 5:1, 21).[12] Here we see Jesus carrying out the mission given to him by his heavenly Father to be "A LIGHT OF REVELATION TO THE GENTILES" (Lk 2:32). Jesus enters into this unclean region contaminated by the presence of Gentile sinners. To make matters worse, he encounters a man "with an unclean spirit" who lives among the unclean dead in the unclean tombs. And it gets worse! Mark tells us that in this unclean Gentile region "there was a large herd of swine feeding nearby on the mountain" (Mk 5:11). Mark shows Jesus and his disciples entering one of the most contagious and ritually contaminated environments an observant Jew could imagine, but contrary to the accepted thinking of traditional Jewish religious leaders, none of this uncleanness contaminated Jesus and his disciples. Instead, Jesus infects this man and the entire region with his holiness. Jesus sends this transformed man back to his own region and people "and he went away and began to proclaim in Decapolis what great things Jesus had done for him; and everyone was amazed" (Mk 5:20). The Greek word Mark uses for "proclaim" (*kēryssō*) is one of the most frequent New Testament words for "gospel witness" or "evangelization."[13] Jesus turned this contaminated, unclean, wild man into an evangelist of the good news of salvation to begin the process of spiritual decontamination in the Gentile region of the Decapolis.

We must note something very important in Mark's narrative of what Jesus does after leaving this highly contaminated environment.[14] From the establishment Jewish perspective, Jesus and his disciples should have gone immediately to the temple to begin the process of ritual purification and decontamination. Instead, they meet a Jewish synagogue official, become contaminated by the woman with a hemorrhage, enter into the home of the Jewish synagogue official, and then travel to Jesus' hometown synagogue in Nazareth, where he teaches on the sabbath (Mk 5:21–6:1).

---

[12]Charles R. Page II, *Jesus and the Land* (Nashville: Abingdon, 1995), 81-91.

[13]Michael Green, *Evangelism in the Early Church* (Grand Rapids: Eerdmans, 1970), 58-66.

[14]Jennifer Wilkinson, "Mark and His Gentile Audience: A Traditio-Historical and Socio-Cultural Investigation of Mk 4.35-9.29 and Its Interface with Gentile Polytheism in the Roman Near East" (PhD diss., Durham University, 2012). She notes that the Decapolis in the first century was known as a center of polytheism, including the worship of Zeus, Artemis, the Nabatean god Pakeidas, and the Roman god Dionysus.

Whereas in the Old Testament God's holiness needed to be quarantined from potential infections, in the teachings of Jesus, holiness is the real contagion. Borg puts it this way: "Holiness—the power of the holy, of the sacred—was understood as a transforming power, not as a power that needed protection through rigorous separation."[15] In the Gospels, contagious holiness seems to describe the infectious quality of Jesus and those who follow him and receive his salvation. Believers who trust in Jesus for salvation receive the Holy Spirit who thus makes them holy ones (*hagioi*, meaning "saints").

CONTAGIOUS GOSPEL

The Christian's connection to Jesus (union with Christ) infuses the believer with Christ's holiness and makes believers spiritually contagious. One can deduce from these examples of contagious holiness in the ministry of Jesus that the proclamation and the living out of the good news of salvation spreads from holy people to sinful people, thereby giving them the opportunity to become holy too.

We observe a similar contagious quality in the spread of the Christian faith in the book of Acts. Luke often uses phrases such as, "The word of God kept on spreading; and the number of the disciples continued to increase greatly in Jerusalem, and a great many of the priests were becoming obedient to the faith" (Acts 6:7). Luke uses summarizing progress reports like this to describe the extent to which the word of the gospel spread as the disciples moved from Jerusalem to Rome (Acts 12:24; 13:49; 19:20).

The word of God, which Luke says is spreading, Paul refers to as "the holy Scriptures" (Rom 1:2). If the word of God is holy, it shares the contagious quality that enables it to "spread" from person to person and from region to region, as disciples proclaim that word, which is indwelt and empowered by the Holy Spirit. In 2 Thessalonians 3:1, the apostle Paul writes to the church in Thessalonica and asks them to "pray for us that the word of the Lord will spread rapidly and be glorified, just as *it did* also with you." The contagious quality of the holy word of God, proclaimed by God's holy ones and empowered by the Holy Spirit, will spread, and spread "rapidly" as Paul desired.

---

[15]Borg, *Conflict*, 147.

Luke closes the book of Acts in a way that suggests the contagion of the gospel continued to spread after the events he recorded. The last word in the Greek text of Acts is *akōlytōs*, an adverb meaning "without hindrance." This word occurs only here in the New Testament, and Luke uses it with singular effect to convey, as Darrell Bock notes, "The gospel is spreading to the entire world."[16]

Rodney Stark picks up the story where Luke leaves off by providing us with a sociological analysis of the spread of the Christian faith in the Roman Empire. He argues that the gospel spread through the Roman Empire because "Christians maintained open networks."[17] That is, Christians passed on their faith through "direct and intimate interpersonal attachments" with those outside the faith.[18] Their social networks were not limited to other Christians but were "open" to non-Christians, thus enabling them to spread the contagion of the gospel from one "interpersonal attachment" to another. Stark concludes that "the primary means of its growth was through the united and motivated efforts of the growing numbers of Christian believers, who invited their friends, relatives, and neighbors to share the 'good news.'"[19] Confirmation of this analysis is provided by one antagonistic to the faith. In Pliny's famous letter to the Emperor Trajan, in which he seeks his advice on how to respond to the spread of the Christian faith, he writes, "For the contagion of this superstition has spread not only through the cities, but also to the villages and in the open country."[20] It is quite ironic that Pliny uses the language of contagion to describe the spread of the gospel since that is the language of the spread of influence in a social network. This social network contagion is exactly what Alan Hirsch describes when he writes, "The gospel . . . travels like a virus. It is 'sneezed' and then passed on through further sneezing from one person to another. All that is needed are the right conditions and the appropriate relationships into which we can 'sneeze.'"[21] The "appropriate relationships" to which he refers are those "interpersonal attachments" Stark observed operating in the open social network of the early church.

---

[16]Darrell L. Bock, *Acts*, Baker Exegetical Commentary on the New Testament (Grand Rapids: Baker Academic, 2007), 759.

[17]Rodney Stark, *The Rise of Christianity* (New York: HarperOne, 1997), 21.

[18]Ibid., 20.

[19]Ibid., 208.

[20]Hans Conzelmann, *History of Primitive Christianity* (Nashville: Abingdon Press, 1973), 169.

[21]Alan Hirsch, *The Forgotten Ways: Reactivating the Missional Church* (Grand Rapids: Brazos Press, 2006), 211.

CONTAGIOUS SANCTIFICATION

Just as the gospel spreads from one person to another through social net-works, so does a Christian's faith spread from one believer to another. As the Spirit of God works in and through the connections and interactions that comprise the spiritual ecology of the church, the holiness that characterizes individual believers and the body of Christ as a whole spreads as a conta-gious spiritual influence from one member to another.

Paul suggests such a concept of holiness in Romans 11:16: "If the first piece *of dough* is holy, the lump [loaf] is also; and if the root is holy, the branches are too." In the context of Romans 9–11, Paul is of course referring to the relationship between Jews and Gentiles and God's permanent commitment to Israel. He uses the allegory of the olive tree to illustrate the relationship between the two groups. Scholars argue over the details of this verse, espe-cially whom Paul understands to represent the "first piece of dough," the "lump [loaf]," the "root," and the "branches."[22] In spite of the disagreements over these textual details, most scholars agree that Paul's thought process moves "from the part to the whole."[23]

We do not want to engage the scholarly debate over this passage. We simply want to call attention to the truth that Paul enunciates in Romans 11:16 regarding the contagious and infectious quality of holiness. Israel is holy and those organically connected to Israel partake of that holiness. Paul's two mixed metaphors—bread and tree—both illustrate "sanctity transference."[24] In particular, he suggests by his use of these two metaphors "something is sanctified when it is added to a sacred entity."[25] The "something" in this passage refers to believing Gentiles who, through Christ, now have a con-nection to "a sacred entity" and now partake of the holiness God bestowed on Israel ("the lump is also" and "the branches are too"). The "sacred entity"—lump and root—is hard to identify specifically but must refer to some manifestation—patriarchs, individual Israelites, Christ, believing

---

[22]The Greek word for this phrase is *aparchē*, most often translated "first fruits" (as in Rom 8:23; 1 Cor 15:20, 23; Jas 1:18; Rev 14:4) and referring to the OT practice of dedicating the first portion of produce, bread, wine, oil, and flocks to God (Num 15:20-21; Deut 18:4).

[23]Douglas J. Moo, *The Epistle to the Romans* (Grand Rapids: Eerdmans, 1996), 698.

[24]Benjamin D. Gordon, "On the Sanctity of Mixtures and Branches: Two Halakic Sayings in Ro-mans 11:16-24," *Journal of Biblical Literature* 135, no. 2 (2016): 363.

[25]Ibid., 358.

Jews—of Israel's status as God's holy people based upon God's gracious choice (Rom 9:14-18). Whatever the "sacred entity" Paul has in mind, the irrefutable fact is that he identifies it as holy and its holiness is transferrable.

Can we not extend the image to the Christian and the church? If one Christian is holy, will not that holiness permeate the entire environment the holy Christian inhabits? If Israel's holiness, which is at issue in this passage, can spread from believing Jews to believing Gentiles, can we not expect a similar dynamic to continue in operation when those same Jews and Gentiles are "being fitted together . . . growing into a holy temple in the Lord"? (Eph 2:21). The holy person permeated with God's holiness is contagious with the potential to spread through the spiritual ecology of the church. While each Christian is holy because of his or her relationship to Jesus, each Christian will gradually manifest this holiness over the course of time (Rom 12:1; Eph 1:4; 1 Thess 4:7). Paul can therefore admonish the believers in Corinth that they should be "perfecting holiness in the fear of God" (2 Cor 7:1). The present tense plural form of "perfecting," (*epitelountes*), matches the present tense plural form of "being transformed," (*metamorphoumetha*) in 2 Corinthians 3:18, both of which are possible only through Spirit empowerment. Paul is not advocating human endeavor toward moral perfection but rather refers to "working out the holiness that has already been granted to those who are part of God's people."[26] Further, the working out of his holiness is not a purely individual experience; Paul uses plural forms. This usage suggests that Paul understands both transforming and perfecting as community endeavors of a collective nature.[27]

We have yet to grasp the significance and impact of what Jesus and Paul taught about the contagious character of God's holiness. Our privatized and individualized notions of spiritual formation have inadvertently deprived us of the rich spiritual ecology we inhabit in Christ. Peter Adam identifies the two contrasting but complimentary approaches to spiritual formation when he writes, "Spirituality is often assumed to be a personal if not private matter, and spiritual growth is assumed to be individual rather than corporate . . . Spirituality that is personal and not corporate is

---

[26]Scott J. Hafemann, *2 Corinthians*, NIV Application Commentary (Grand Rapids: Zondervan, 2000), 287.

[27]James M. Howard, *Paul, the Community, and Progressive Sanctification* (New York: Peter Lang, 2007); James G. Samra, *Being Conformed to Christ in Community* (London: T&T Clark, 2006).

sure to be unfruitful, because maturity is fundamentally corporate in the New Testament."[28]

Paul explains how each connected member of the body spiritually influences other connected members through contagious contact. In Ephesians 4:1-16, the apostle describes the concept of the body of Christ beginning with the critical role of the Holy Spirit, who is the unifying glue that creates unity in the midst of the ecological diversity of the body (Eph 4:3-4).

According to Paul, there is "one body and one Spirit" (Eph 4:4). This abrupt phrase indicates that in Paul's mind, the presence of the "one Spirit" makes possible the unity of the "one body." It is the Spirit, then, who makes possible the connection between the head and the rest of the body as well as the connections between the individual members (Eph 4:15-16).[29] Paul uses two *syn*-compounds to express how individual members connect to each other. These spiritual connections, made possible by the joint ministry of Jesus and the Spirit, spread spiritual contagion (*haphē*) throughout the whole body. Harold Hoehner provides ample evidence from both the Old Testament and classical Greek that this word *haphē* "depicts the spread of disease" through physical contact.[30] Markus Barth offers several alternative meanings for *haphē* but concludes that "The etymology of *haphē* suggests the translation, "touch," contact," or "grip."[31] He offers the translation "contact of supply" because it best captures Paul's point "that in their mutual dependence and communication all church members are chosen tools of the head for communicating nourishment, vitality, unity, solidity to the body as a whole."[32]

Based on his analysis of this word, Hoehner concludes that "when applied to the body of believers it is clear that the union and growth of the body can only come when there is contact with other members of the body . . . the interaction or contact between members . . . promotes both individual and

---

[28]Peter Adam, *Hearing God's Word: Exploring Biblical Spirituality* (Downers Grove, IL: InterVarsity Press, 2004), 175-76.

[29]See chapters 9, 10, and 11 for a full treatment of the significance of these Pauline *syn*-compounds.

[30]Harold W. Hoehner, *Ephesians: An Exegetical Commentary* (Grand Rapids: Baker Academic, 2002), 571.

[31]Markus Barth, *Ephesians: Translation and Commentary on Chapters 4–6* (New York: Doubleday, 1974), 449.

[32]Ibid.

corporate growth."[33] In addition to these two proposals, Montague offers additional evidence for the contact/contagion interpretation of *haphē*:

> What then is the theological bearing of the phrase, "growing more and more compact and knit together through every help-ministering contact"? Is this contact with Christ or with fellow-members? The context stresses the role of the members of the body's self-building. The compound prefixes suv- and suµ- confirm this . . . This does not diminish the role of Christ; all growth is *from* him (eξ); but it comes *through* (δια) the unification, mutual contact, interaction and help of the members.[34]

He quotes B. Wescott in this regard, who writes, "Wherever one part comes into close connection with another, it communicates that which it has to give."[35]

What Hoehner, Barth, and Montague seem to suggest is that the reciprocal interactions between interconnected members of the body spread spiritual nutrients in a contagious manner, producing mutual growth throughout the entire ecosystem of the body. In Ephesians 4:16, Paul uses a synonym of *allēlōn* (one another) in the reciprocal pronoun *heautou* (each other) to describe how the body builds itself up *oikodomē* (in love). Reciprocal interactions among connected members produce reciprocal edification or spiritual growth. Thus Montague can say it is "the whole Christian community, which progresses through mutual interaction toward a perfect temple (Eph 2:21)."[36] We should not overlook the fact that the root of the Greek word for "edification" or "building up" is *oikos*, from which we derive our English word "ecology." Again Hoehner astutely observes, "But as in verse 12, the building here is not inanimate but a living and growing organism made up of living believers."[37] This means we must not think of *oikodomeō* as the building up of a physical structure but as the cultivation of a living and growing organism. The cultivating work is the active ministry of the Spirit through the head who makes possible contagious connections between the members of the body, resulting in mutual growth.

---

[33]Hoehner, *Ephesians*, 573.
[34]George T. Montague, *Growth in Christ: A Study in Paul's Theology of Progress* (Kirkwood, MO: Maryhurst Press, 1961), 159.
[35]Ibid., 158.
[36]Ibid., 161.
[37]Hoehner, *Ephesians*, 578.

ECOLOGICAL SANCTIFICATION

We think the infectious nature of holiness is missing from traditional Christian teaching about how we grow in faith and become progressively more holy. A privatized understanding of the ministry of the Holy Spirit and an individualized view of faith formation often hinders Christian understanding of this basic tenet of the teaching of Jesus and Paul.

When we contemplate the ministry of the Holy Spirit in spiritual growth, we often visualize a pipeline between God and ourselves. We see ourselves engaged in a private relationship with God in Christ through the Holy Spirit. We cultivate this relationship through private spiritual rituals, such as solitary Bible reading, solitary prayer, solitary meditation, and solitary worship. There is almost a medieval monastic isolationist quality in this approach to Christian growth. While there is certainly a private dimension to our relationship with God and to our walk of faith, it does not begin and end there. We need to see our personal faith and personal relationship situated within a larger ecology of faith that includes our brothers and sisters in Christ.

Paul's teaching about the relationship between individual members and the whole body of Christ must profoundly inform our perceptions. The interconnected nature of the relationship between members, head, and whole body should have kept us from such a privatized view of faith formation. Unfortunately, we have read these body texts in Paul to inform our doctrine of the church more than to develop our doctrine of sanctification.

We often delimit progressive holiness to a personal and private work of the Spirit in each believer. While there is one aspect of progressive holiness that partakes of this personal, ongoing manifestation of God's holiness, we have often left the discussion there. We think that the personal concept of progressive holiness needs expanding to include the role other believers play in forging both individual and corporate holiness. Phil Needham proposes that we think of holiness "as a function of community."[38] He insists that "a private holiness will not make us caring of one another and builders of community."[39] He suggests that a distorted, privatized concept of personal holiness is due in part to "the lack of a Scripturally

---

[38]Phil Needham, "Integrating Holiness and Community: The Task of an Evolving Salvation Army," *Word and Deed* 3, no. 1 (2000): 7.
[39]Ibid., 8.

based understanding of the Body of Christ."[40] An ecological view of the interconnective and interactive aspects of the body imagery helps us broaden our concept of holiness from individual to corporate. We are right to think in terms of "the social ecology of holiness" to describe the social-spiritual environment where Christians nurture holiness through their interactions with one another.[41] Frederick Coutts recognizes the social dimension of holiness when he writes, "The truth is that a genuine experience of holiness can be developed only in the field of personal relationships . . . The life of holiness cannot flourish in isolation."[42]

We think of relational holiness as ecological sanctification that occurs within the body of Christ as individual members who possess the Holy Spirit connect to and interact with one another in many ways. Ecological sanctification happens when the contagion of holiness spreads through connected and interacting members of the body of Christ. Anthony Hoekema says, "We are sanctified through fellowship with those who are in Christ with us."[43] He elaborates, saying, "Fellowship with other Christians could, in fact, be thought of as one of the more important means of sanctification."[44] Fellowship (*koinōnia*) in the New Testament is the bond of connection we have with Christ and with one another that makes us one (1 Cor 1:9; Phil 2:1; 1 Jn 1:3-7). Fellowship between Christians takes the connections and interactions we have with one another and makes them conduits for the spiritual contagion of holiness. The more frequent our relational contacts with our brothers and sisters in Christ, the more opportunities we have to spread the contagion of holiness throughout the body of Christ.

As we have seen, Paul viewed the church as a "holy temple in the Lord." However, the church, like individual Christians, must become holy by gradually "lay[ing] aside the old self" and gradually "put[ting] on the new self" (Eph 4:22, 24). To use the language of Irenaeus, "Christ became what we are, in order that we might become what he is."[45] According to Paul, God "made

---

[40]Ibid., 12.

[41]Jonathan S. Raymond, "Social Holiness: Journey, Exposures, Encounters," in *The Holiness Manifesto*, ed. Kevin W. Mannoia and Don Thorsen (Grand Rapids: Eerdmans, 2008), 175.

[42]Frederick Coutts, *The Splendor of Holiness* (London: Salvationist Publishing, 1983), 16-17.

[43]Anthony A. Hoekema, *Saved by Grace* (Grand Rapids: Eerdmans, 1989), 195.

[44]Ibid., 195n4.

[45]Saint Irenaeus of Lyons, *Against Heresies*, ed. Alexander Roberts and James Donaldson (South Bend, IN: Ex Fontibus Company, 2015).

Him who knew no sin *to be* sin on our behalf, so that we might become the righteousness of God in Him" (2 Cor 5:21). Becoming what he is requires a process of sanctification or spiritual maturity that gradually transforms us so that one day we will stand "complete in Christ" (Col 1:28). That process is not just an individual one carried out through solitary spiritual exercises but is also a corporate one carried out through relational interactions (including corporate worship) with other holy believers who comprise "the fellowship of the Holy Spirit" (2 Cor 13:14).

The Holy Spirit does not simply work in individual Christian hearts to sanctify and make us complete. Numerous passages in the New Testament display the Holy Spirit carrying out his ministry through the corporate community of believers. Jesus himself sets the stage for our understanding of the Spirit's work in the community in John 16:13: "But when He, the Spirit of truth, comes, He will guide you into all the truth; for He will not speak on His own initiative, but whatever He hears, He will speak; and He will disclose to you what is to come." Both of the personal pronouns translated "you" are in plural form in the original Greek text. Here Jesus promised his original and subsequent disciples that the "Spirit of truth" would guide and disclose to the community of believers, not just to a select few. The promise is for corporate guidance, not individual direction.

Several excellent works describe the Spirit's work in and through the church community leading to individual and corporate growth and maturity.[46] This much-needed emphasis counterbalances the highly individualistic treatment found in most discussions of the Spirit's ministry in the church. For instance, Howard states, "It is the work of the Spirit in the community that makes possible the growth to maturity . . . the basic definition of progressive sanctification."[47] Taking a slightly different approach, but reaching the same conclusion, Samra writes, "By facilitating the components of the maturation process, which are the means the Spirit uses to conform believers to Christ, the believing community is able to facilitate the process of maturation. Thus, believers are being conformed to Christ in community."[48]

---

[46]Howard, *Paul, the Community, and Progressive Sanctification*; Samra, *Being Conformed*; Karl Barth, "The Holy Spirit and the Gathering of the Christian Community," in *Church Dogmatics*, IV/1, 643-739; Montague, *Growth in Christ*, 144-64.

[47]Howard, *Paul, the Community, and Progressive Sanctification*, 185.

[48]Samra, *Being Conformed*, 166.

When we read Acts 2:1-4 through our Western, individualistic eyes, we miss the community emphasis Luke wants us to see. Notice that "all" the disciples were in the room and the tongues of fire rested on "each one" of the disciples and that "all" of the disciples were filled with the Holy Spirit and "all" began to speak in other tongues as the Spirit empowered "them." The filling of the Spirit as empowerment for mission is a corporate experience, not an individual one! We see a similar experience of corporate filling in other places in the book of Acts (see Acts 4:31; 10:44; 11:15; 13:52; 19:6). This does not rule out individual filling of the Spirit (see Acts 4:8), but it reminds us that both are legitimate expressions of the way in which the Spirit works in the spiritual ecology of the body of Christ. We never want to sacrifice one for the other, but we must recognize that both are necessary manifestations of the Spirit's work in God's people.

REFRAMING SANCTIFICATION

Thinking about sanctification through an ecological lens means we have to reframe some of our traditional thinking about how Christians grow and how the Holy Spirit ministers in and through Christians embedded in a spiritual ecology. An ecological frame of reference alerts us to certain dynamic qualities of ecosystem functions that prevent us from thinking in purely personal categories about spiritual growth.

Thinking ecologically about sanctification also means we have to think ecologically about sin. Holiness and sin—impurity, defilement—both possess a contagious character that spreads through social and spiritual ecologies. We see this in two cases—one from the Old Testament and one from the New. In Joshua 7, Achan alone violated the ban God imposed on taking the spoils of war in battle against Israel's enemy. His individual sin was so contagious that it infected the entire nation and exposed them to God's wrath and retribution. The only solution was the elimination of the contamination so that the nation could be spared of God's judgment. In 1 Corinthians 5, Paul confronts the sin of sexual immorality of an unnamed "someone" who took his father's wife as his own (1 Cor 5:1). Instead of mourning over this violation of God's law, some in the church were "boasting" because they did not realize, as Paul did, that "a little leaven leavens the whole lump *of dough*" (1 Cor 5:2, 6). The only solution to this contagious influence was to "clean out the leaven" before it spread further and contaminated more individuals and eventually contaminated the whole church ("the whole lump of dough").

The Corinthian text confronts us with an important question based upon Paul's actions with the church in Corinth. Why didn't Paul advocate a positive influx of holiness to overcome the power of the contaminating sin? One may find the answer to this question in the spiritual condition of the church at that time. The spiritual ecology of the church in Corinth was weak through the deliberate actions and attitudes of many of its members. Paul describes them as "divided," "men of flesh," "infants in Christ," "fleshly," "acting like mere humans," "arrogant," immoral, "have lawsuits with one another," "contentious," and "drunk" (1 Cor 1:10; 3:1, 3 NIV; 4:6, 18-19; 5:1, 6:7; 11:16, 18-19, 21). They as a local congregation are not in any spiritual condition to offer an effective positive resistance to the infection of sin. Their spiritual immune system is so weak that the only solution for them at the time is to remove the infection so that the body can begin the process of spiritual renewal and restoration. Paul tells them plainly in his critique of their celebrations of the Lord's Supper, "Many among you are weak and sick" (1 Cor 11:30). This was apparently true both physically and spiritually.

Paul's solution to this state of spiritual immune deficiency is to strengthen the body. Thus, he calls upon them to "never eat meat again" if it causes a brother or sister to stumble, "let no one seek his own *good*, but that of his neighbor," and "give no offense . . . to the church of God" in order to seek "the *profit* of the many" (1 Cor 8:13; 10:24, 32-33). He also calls upon them to "wait for one another," "have the same care for one another," "pursue love," "seek to abound for the edification of the church," and to "greet one another with a holy kiss" (1 Cor 11:33; 12:25; 14:1, 12; 16:20).

All of these apostolic admonitions direct God's people to "one another" (*allēlōn*) relationships designed to strengthen ecological connections and interactions in the body of Christ. The church in Corinth was "divided" and disconnected as a spiritual ecology. Social, economic, and ethnic divisions disrupted the ecological flow of spiritual nutrients from one member to another, contrary to what God intended (1 Cor 12). As long as this condition continued, the church in Corinth could not function as a spiritual ecology that edified itself through the connections and interactions that made them one body. This explains why Paul is so adamant about the problem of divisions and factions that plagued the congregation. Some members were spiritually immature and infantile because they had disconnected from each other and had thereby cut themselves off from each other's spiritual nourishment. All organisms in ecologies grow by sharing nutrients and resources through connections and interactions that constitute the multiple relationships between and among them. When an internal or external threat disrupts these connections and interactions, it interrupts the reciprocal flow of nutrients and renders individual organisms and the entire ecology vulnerable and weak. Such was the condition of the spiritual ecology of the church in Corinth. Paul as the one who planted and Apollos as the one who watered the field in Corinth admonished them to reconnect to and interact with one another in order to repair the broken ecosystem and make it function again as God intended (1 Cor 3:6-9).

It often seems that our traditional Christian views of holiness and sin are closer to those of traditional Second Temple Judaism than they are to the teachings of Jesus and Paul. Many Christians think holiness is fragile and sin is powerful, and so Christians need to avoid the contamination of sin by avoiding sinners and sinful culture. Many "monastic Christians" withdraw from social or cultural contact with the non-Christian world (society and culture) because they fear contamination and defilement and want to protect their fragile holiness. This may be what Rod Dreher is getting at when he writes, "Communities that are wrapped too tight for fear of impurity will suffocate their members and strangle the joy out of life together."[49] Such Christian communities have a defective view of the power of Christ's holiness to protect them and to overpower sin. We forget those words of John the apostle in 1 John 4:4: "Greater is He who in you than he who is in the world." This in spite of the fact, as John also writes, that "we know that we are of God, and that the whole world lies in *the power of* the evil one" (1 Jn 5:19). The power of Christ manifested in the empowering presence of the Holy Spirit is so strong that John writes, "We know that . . . the evil one does not touch" the one who is "born of God" (1 Jn 5:18). This truth and power emboldened and enabled Jesus to break missional barriers,

---

[49]Rod Dreher, *The Benedict Option: A Strategy for Christians in a Post-Christian Nation* (New York: Penguin Random House, 2017), 139.

as we saw in Mark 5, confront the power of Satan in the demon-possessed man, and overpower Satan and his demons. If we retain a fear of the power of sin and Satan, we will never achieve the mandate of the Great Commission to "make disciples of all the nations" (Mt 28:19). Jesus gave an example in Mark 5 and in other events of his mission from God that demonstrate the power of his holiness in the face of sinful and satanic opposition that operates in those nations.

CONTAGIOUS HOLINESS IN THE ONLINE CLASSROOM

The transfer from condemned sinner to delivered saint is what theologians mean by justification. Justification means God declares us righteous and holy in his eyes. The apostle Paul refers to this radical change in 2 Corinthians 5:17 (NIV): "Therefore, if anyone is in Christ, the new creation has come: The old has gone, the new is here!" At the moment of transfer from the kingdom of darkness to the kingdom of light, we are not only justified (made righteous) but sanctified (made holy), but our status as saints or holy people must progressively manifest itself through a process of spiritual maturity. The process of becoming holy or sanctified is both an individual and corporate one. It is a process empowered by the Holy Spirit, working individually and collectively, to extend and manifest God's holiness in God's people. The Holy Spirit enables Christians to spread holiness from one person to another through our connections to one another as members of the body of Christ. Through various kinds of engagements, interactions, and transactions, Christians spread the contagion of holiness among themselves. One of the mechanisms God provides for the spread of holiness among his people is the exchange of spiritual nutrients and resources through living out the "one another" imperatives found one hundred times in the New Testament. We also spread the contagion of holiness among ourselves through the preaching and teaching of God's word, intercessory prayer, the sharing of the Lord's Supper, interactive experiences of worship, learning together both in person and online, and joint enterprises of service in which we labor together to accomplish the mission of the church in our neighborhoods and, ultimately, our world.

Our faith makes us contagious to non-Christians but our holiness makes us contagious to one another. When we form online communities of faith in courses and degree programs we are creating opportunities through digital connections and technological interactions to spread our spiritual contagion to each other. We can "one another" our brothers and sisters whether we are in a classroom or in an online course. We can encourage one another, pray for one another, and teach one another even though separated by time and space. We have seen it happen and we have been both givers and receivers of this sharing with students we have never met in person. It is a testimony to the power of the Spirit who works through the connections and interactions we create online to form us as a community into the image of God's Son.

In the final chapter, we provide a comprehensive summary of the salient features of our ecologies of faith model, which will enable you to reflect upon the previous chapters and consider possible applications to your own ministry context.

— THIRTEEN —

THINKING ECOLOGICALLY ABOUT SPIRITUAL GROWTH

*[The] biblical vision offers us an ecological understanding of creation
and new creation that is closely tied to the work of Jesus Christ
and the biblical narrative of the kingdom of God.*

HOWARD A. SNYDER,
SALVATION MEANS CREATION HEALED

Jacqueline Mattis et al. write, "It is critical that scholars take an ecological approach to studies of spiritual maturity."[1] We found this comment buried in a five hundred-page scholarly publication on the subject of the spiritual development of children and youth. While the book is widely acclaimed and regularly referenced, the ecological perspective on spiritual growth it encourages is not a dominant orientation of researchers, professors, youth leaders, or pastors. It seems we think about, conceptualize, and practice spiritual formation from a default position influenced by theological and denominational traditions. While natural and social scientists are in the process of forging a science of networks and ecosystems, the Christian community has yet to formulate a robust theology of networks and ecosystems. Part of such a theology includes an ecological understanding and appreciation for how Christians grow spiritually.[2]

---

[1]Jacqueline S. Mattis et al., "Ethnicity, Culture, and Spiritual Development," in *The Handbook of Spiritual Development in Childhood and Adolescence*, ed. Eugene C. Roehlkepartain et al. (San Francisco: Sage Publications, 2006), 293.
[2]Heidi Campbell and Stephen Garner, *Networked Theology: Negotiating Faith in Digital Culture* (Grand Rapids: Baker Academic, 2016).

Whether or not we are aware of it, we have had our perspectives and perceptions of Christian spiritual maturity influenced more by Western individualism than biblical individuality.[3] Western individualism glorifies and exalts in the autonomous individual who forges his or her own destiny in the world. Iconic images of the American pioneer, rugged cowboy, and even lone inventors, such as Thomas Edison, Nikola Tesla, George Washington Carver, and Grace Hopper, dominate the North American cultural landscape. Iconic images are often distortions of reality. American pioneers rarely traveled alone; they were with others in a wagon train. Cowboys such as the Lone Ranger and Roy Rogers all had sidekicks or companions who assisted them in fighting the bad guys. Inventors like Tesla and Edison did not discover new inventions and innovations while working alone in their laboratories. Instead, as Christopher Cooper writes, "The myth of this lone-genius inventor . . . [ignores] the true collaborative and incremental process of invention."[4]

Biblical individuality recognizes the existence and importance of the individual but always contextualizes and situates the individual within some larger human community. The Scriptures give a rightful place to individuals God used to move his missional agenda for the world forward, culminating in the man Christ Jesus.[5] Biblical individuality never obliterates the individual person in favor of the group. On the day of Pentecost, recorded in Acts 2, we see individual believers filled with the Spirit and hearing the good news of salvation in their own native languages ("each one of them," "we each hear," Acts 2:6, 8). In Paul's description of spiritual gifts in 1 Corinthians 12:11, he notes that the "Spirit works all these things, distributing to each one individually just as He wills." In that fantastic scene John describes in Revelation 7:9, he finds those "from every nation and *all* tribes and peoples and tongues, standing before the throne and before the Lamb." While this multitude comprises a unified group, no one surrenders his or her individual ethnicity

---

[3]C. Norman Kraus, *The Authentic Witness* (Grand Rapids: Eerdmans, 1979); Miroslav Volf, *Exclusion and Embrace: A Theological Exploration of Identity, Otherness, and Reconciliation* (Nashville: Abingdon Press, 1998); Bruce J. Malina, *The New Testament World: Insights from Cultural Anthropology* (Louisville: Westminster John Knox, 2001), 58-80; E. Randolph Richards and Brandon J. O'Brien, *Misreading Scripture with Western Eyes* (Downers Grove, IL: InterVarsity Press, 2012).
[4]Christopher Cooper, *The Truth About Tesla: The Myth of the Lone Genius in the History of Innovation* (New York: Race Point, 2015), jacket cover.
[5]Avery T. Willis Jr. and Henry T. Blackaby, *On Mission with God: Living God's Purpose for His Glory* (Nashville: Broadman & Holman, 2002).

because John can still recognize the differences and, indeed, celebrates them. However, these individuals always have some relationship to a larger group or community. Israel's kings were individuals who represented the entire nation and the ultimate King Jesus as Israel's Messiah.[6] As N. T. Wright says, "*Christos* denotes 'the Messiah and his people,' or perhaps better 'the Messiah as the representative of his people,' the one *in whom* that people are summed up and drawn together."[7] The Old Testament concept of "corporate personality," although in need of nuance and modification from its original description, still illuminates a biblical understanding of the undisputed fact that Hebrew thinking, reflected in the Old Testament Scriptures, viewed the individual as reciprocally engaged within a larger social unit—family, clan, tribe, nation.[8] Pauline scholars recognize the concept of corporate personality as exerting some influence on Paul's Adam-Christ imagery as well as his concept of the body of Christ.[9] Adam and Christ are Paul's reflections on the relationship between the individual and the group that pertain to the person and work of Christ and his relationship to those who believe in him. An ecological perspective on spiritual maturity illustrates the need for a balance between individual organisms and the larger ecosystem in which they grow and flourish. Ecologist Walter Taylor describes this balance:

> The individual organism is sometimes thought of as one capable of independent existence. But few or no organisms are capable of independent existence. They depend utterly, in final analysis, on other individuals or species. . . . Apparently there is little rugged individualism in nature.[10]

---

[6]Aubrey R. Johnson, *Sacral Kingship in Ancient Israel* (Cardiff: University of Wales Press, 1967); S. Mowinckel, *He That Cometh: The Messiah Concept in the Old Testament and Later Judaism* (Uppsala: Almqvist & Wiksells, 1943).

[7]N. T. Wright, *Paul and the Faithfulness of God* (Minneapolis: Fortress, 2013), 834.

[8]J. Rogerson, "The Hebrew Conception of Corporate Personality: A Re-examination," *Journal of Theological Studies* 21 (1970): 1-16; Andrew Perriman, "The Corporate Christ: Reassessing the Jewish Background," *Tyndale Bulletin* 50, no. 2 (1999): 239-63; H. Wheeler Robinson, *The Christian Doctrine of Man* (Edinburgh: T&T Clark, 1934); *Religious Ideas of the Old Testament* (London: Duckworth, 1949); *Corporate Personality in Ancient Israel* (Philadelphia: Fortress, 1980); Aubrey R. Johnson, *The Vitality of the Individual in the Thought of Ancient Israel* (Cardiff: University of Wales Press, 1949); Johs. Pedersen, *Israel: Its Life and Culture*, vols. 1-2 (London: Geoffrey Cumberlege, 1964).

[9]Ernest Best, *One Body in Christ* (London: SPCK, 1955); Jean De Fraine, *Adam and the Family of Man* (Staten Island, NY: Alba House, 1965); James D. G. Dunn, *The Theology of Paul the Apostle* (Grand Rapids: Eerdmans, 1998).

[10]Walter P. Taylor, "Significance of the Biotic Community in Ecological Studies," *Quarterly Review of Biology* 10, no. 3 (1935): 296.

Individual spiritual growth is never exclusively individual; it always plays out in some larger spiritual ecology as individual Christians connect to and interact with the Godhead, God's word, God's Spirit, and other members of the body of Christ.

We have argued that thinking biblically about spiritual growth involves thinking ecologically since the Scriptures point us in that direction.[11] We have seen that the Bible is an ecological book in the sense that it regularly refers to ecological realities in God's original creation and in his new creation in Christ. Biblical writers consistently draw upon original creation realities to explore and explain similar realities that operate in the kingdom, the church, and the Christian. Alan Hirsch says that "we are on fertile biblical ground" when we use organic and ecological language to describe the church and kingdom.[12] The nature psalms of the Old Testament and the nature parables of Jesus draw upon these natural realities in ways we are just now beginning to appreciate. Those who lived during biblical times lived close to the land in rural settings with close ties to agriculture, horticulture, and viticulture. They more easily grasped ecological realities than we who live in urban centers in the twenty-first century. It has taken the scientific study of ecology, systems, and networks to arrest our attention and convince us of their existence and value. All the while, the Scriptures remained available to us, but we did not read or interpret them through ecological eyes, and we regularly missed a very basic orientation to the subject of spiritual formation. Thinking ecologically about spiritual growth involves a number of propositions that collectively offer convincing evidence for reframing spiritual growth from an ecological perspective. When we reframe a painting, we change how we view it. When we reframe an experience, it changes how we perceive that experience. When we reframe a concept, such as spiritual formation, it changes how we perceive that concept. By mounting spiritual growth within a new frame of reference, we are able to think differently about it and consider a new perspective. Think of the following propositions

---

[11]Howard A. Snyder, *Liberating the Church: The Ecology of Church and Kingdom* (Downers Grove, IL: InterVarsity Press, 1983); *Decoding the Church: Mapping the DNA of Christ's Body* (Grand Rapids: Baker Books, 2002); *Salvation Means Creation Healed: The Ecology of Sin and Grace* (Eugene, OR: Cascade Books, 2011); Alan Hirsch, *The Forgotten Ways: Reactivating the Missional Church* (Grand Rapids: Brazos Press, 2006), 179-216.
[12]Hirsch, *Forgotten Ways*, 180.

as summary statements of what we have laid out at length, which comprise the framing structure of our ecologies of faith concept.

Proposition 1: God created a universe that exists and functions as a cosmic ecosystem. Both the testimony of Scripture and evidence from scientific observation confirm the ecological character of God's creative masterpiece. Howard Snyder notes that "the biblical worldstory is at heart ecological. . . . It shows how every part of creation is connected with every other."[13] We noted in chapter one that scientists are now conceptualizing the entire universe as one enormous ecosystem. Edward Wemberley writes, "The planet is literally bathed in cosmic radiation not only from the sun, but from the depths of the universe itself."[14] While scientists explore the far reaches of this "cosmic ecology," some, such as Wemberley, recognize that "there are limitations to how far or how effectively reason and science can be applied to understanding our cosmic 'household.'"[15] He insists that theologians and religious leaders provide explanatory tools not available to scientists that "[extend] scientific formulations of our cosmic ecology" into what he calls "the ecology of the unknown."[16] The Christian community stands at a significant moment in the current state of scientific knowledge about the ecological interconnectedness of God's universe. We have the unique opportunity to engage the scientific community with the inspired perspective of God's word and its implications for how we can experience the final reconciliation of all things in Christ (Eph 1:10).

Proposition 2: The earth exists within a larger cosmic ecology and operates by ecological laws. Gibson offers evidence that the entire earth operates according to the laws of what he calls "ecological mechanics."[17] He defines this as "a change of layout due to complex forces" that results in "any alteration of the shape of the surface of the environment."[18] As examples, he cites things like the rippling and pouring of water, the changes of rubber and clay, and the rupturing of a surface as occurs in an earthquake. When

---

[13]Snyder, *Creation Healed*, 139.

[14]Edward T. Wimberley, *Nested Ecology: The Place of Humans in the Ecological Hierarchy* (Baltimore: Johns Hopkins University Press, 2009), 67.

[15]Ibid., 77.

[16]Ibid.

[17]James J. Gibson, *The Ecological Approach to Visual Perception* (New York: Psychology Press, 2015), 88.

[18]Ibid., 87.

we were in Mary's adopted homeland of Haiti during the 2010 earthquake that devastated the country, we saw these ecological mechanics firsthand. We observed large chunks of the earth's surface altered, and everything sitting on it was pulled apart with it. These changes in the ecological structure of the earth involves going and coming so that "any rigid motion of a body . . . has an equivalent motion in the opposite direction" because all of these natural entities are interconnected and exert mutual influence.[19] We see an equivalent reality in the spiritual ecology of the body of Christ. The spiritual entities represented by individual members of the body of Christ interconnect with Christ, the head, and with each other. As such, their interconnectedness makes possible the exertion of mutual spiritual influence as they engage and interact with one another. The same ecological principle applies when Paul changes his metaphor from body to building "because carpenters and builders are familiar with this branch of physics."[20] The interconnected structure of a building operates under the same ecological mechanics as the physical structure of the earth and the skeletal structure of the human body. We not only *see* this in our own houses but we can *hear* it as well. We can sit in our living room and hear the wooden frame of the house creak in response to some pressure exerted from the wind or the weight of our feet on the wooden floor. When we sit in one particular chair in our house, we can hear a creaking sound from the other side of the room. Maybe we need to lose some weight, or maybe even a slight modification in the ecological mechanics of our house in one spot influences another. Isn't this the point Paul makes regarding our mutual spiritual influence on one another?

Proposition 3: Natural growth follows ecological laws and teaches us that everything grows through ecological interconnections and organic interactions in a mutualistic relationship of interdependence. Growth is a natural byproduct of connections and interactions in niches, habitats, communities, defined ecosystems, and the ecosphere. Consequently, Ernest Callenbach asserts, "Strictly speaking, there are no truly individual organisms. Nothing alive exists in isolation from its ecological context. . . . Symbiotic relationships . . . are the universal way in which life forms survive and coexist."[21] Or, to state

---

[19]Ibid., 89.
[20]Ibid.
[21]Ernest Callenbach, *Ecology: A Pocket Guide* (Berkeley: University of California Press, 2008).

it differently, "The behavior of every living member of the ecosystem depends on the behavior of many others."[22] We ought to ponder the astute observation that "there is as much need today for a Declaration of Interdependence as there was a need for a Declaration of Independence in 1776."[23] Ecological interdependence focuses our attention on relationships within a defined ecosystem, whether that ecosystem consists of natural or spiritual entities. This is the reason why Paul and John, each in their own way, stress the importance of maintaining functioning interpersonal relationships in the spiritual ecology of the church community. Paul writes, "I urge Euodia and I urge Synthyche to live in harmony in the Lord" (Phil 4:2). John writes, "Now I ask you, lady, not as though I *were* writing to you a new commandment, but the one which we have had from the beginning, that we love one another" (2 Jn 5). Paul and John were both realists who recognized that even Christians who were "in Christ" and enjoyed the "fellowship of the Spirit" would occasionally irritate or offend one another. In both Corinthians and Philippians, Paul saw the spiritual threat those broken relationships posed to the proper functioning of the spiritual ecology of the church. Interdependence only works when the ties that connect the different entities in an ecosystem operate properly by exchanging nutrients and resources mutually. When we damage or sever spiritual exchange mechanisms with interpersonal conflict, we disrupt spiritual nurture.

We see this principle indirectly referenced in 1 Peter 2:1-2. We are often so enamored with the word picture Peter paints in verse two that we overlook the prelude to his picture in verse one: "Therefore, putting aside all malice and all deceit and hypocrisy and envy and all slander." The verb "put away" (*apothemenoi*) is an aorist participle indicating that this action is to occur before the action of the verb "long for" (*epipothēsate*) in verse 2. All of the terms Peter uses here are indicative of social sins or sins that disrupt interpersonal relationships and Christian fellowship. Malice (*kakia*) has to do with the intention to injure another person through words or actions. Deceit (*dolos*) is any attempt to trick or fool another person into believing what is

---

[22]Fritjof Capra, *The Web of Life: A New Scientific Understanding of Living Systems* (New York: Anchor Books, 1996), 298.

[23]Donald Worster, *Nature's Economy: A History of Ecological Ideas*, 2nd ed. (New York: Cambridge University Press, 1994), 320.

not true. L. Oberlinner describes the term as behaviors that "trouble the relationships between individuals."[24] Hypocrisy (*hypokrisis*) is pretense or hiding one's identity or motives from another person. Envy (*phthonos*) is a disposition of displeasure at the success or good fortune of another. Slander (*katalalia*) is willful spoken or written injury of another person. All five vices, when practiced in the church, disrupt social harmony. If we are to grow like newborn babies, we must first rid ourselves of these interpersonal sins. If we retain the social vices in the church itemized in 1 Peter 2:1, we will never be able to experience the spiritual growth described in 1 Peter 2:2.

Proposition 4: Ecological laws that govern natural growth operate similarly in the spiritual realm. This principle of spiritual ecology finds its clearest presentation in the nature parables or the parables of growth. New Testament scholars have identified several of the parables Jesus taught as nature parables or parables of growth because they identify a feature of natural growth to illuminate some aspect of the kingdom. The growth of plants, seeds, and flowers—which receive the primary attention from Jesus—result from their interaction with the surrounding environment—air, soil, sun, other species, etc. These interactions are the result of various types of organic connections each plant, seed, or flower has to other biotic and abiotic elements in the ecosystem. The interactions refer to exchanges of nutrients and other necessary resources to sustain life and produce varying degrees and types of growth. All things being equal in an ecosystem—characterized by equilibrium—the connected plants, seeds, and flowers will grow to their maximum capacity. Internal and external threats—perturbation, disturbances, disruptions, and stress—will affect both the capacity for and the rate of growth. Biodiversity is the strongest deterrent to such threats because diverse ecosystems "contain more species in complex interactions and thus can more easily refill niches if their former inhabitants disappear."[25]

The ecological realities of growth, which Jesus points to as illustrations of and points of comparison to the growth of the kingdom, reached its climax in him. We noted in chapter three the insightful comment of C. H. Dodd,

---

[24]L. Oberlinner, "*dólos*," in *Exegetical Dictionary of the New Testament*, ed. Horst Balz and Gerhard Schneider (Grand Rapids: Eerdmans, 1999), 1:344.
[25]Callenbach, *Ecology*, 41.

who sees "no mere analogy, but an inward affinity, between the natural order and the spiritual order."[26] We also quoted, with approval, A. M. Hunter's maxim that the parables Jesus told were "designed to illuminate some spiritual truth, on the assumption that what is valid in one sphere is valid also in the other."[27] In an earlier work, Hunter argued that the "major premise" of our Lord's use of parables was that "nature" was a "revelation of God and his ways" and that Jesus used this revelation "in the service of his heavenly truth."[28] Dodd seems to concur with Hunter, writing, "Since nature and super-nature are one order, you can take any part of that order and find in it illumination for other parts."[29] This is why Dodd can say, "The Kingdom of God is intrinsically *like* the processes of nature and of the daily life of men."[30] Thus, both parable scholars suggest that parables were actual observations of the similarity of the processes of growth viewed as "one order" by our Lord.[31]

We may speculate as to the reason for this consistent pattern in Jesus' nature parables. It may be that Jesus follows this movement from the known processes of growth in nature to the unknown processes of growth in the kingdom because the unknown was also unobservable. While we might be able to observe *that* growth occurred, we might not be able to observe *how* the growth occurred. Jesus is illuminating *how* spiritual growth occurs in the kingdom, and he appeals to the natural growth of plants, flowers, and seeds to provide us with that information. We know from Paul's comment in 1 Corinthians 3:7 that the source and energy for this growth comes from God. This may also explain why Jesus directs his disciples to "observe *how* the lilies of the field grow" (Mt 6:28, emphasis ours). The observation of natural growth provides needed information to the disciples, and us by extension, about how growth occurs in the kingdom. Why is this information necessary? If we and the disciples are to guide, assist, and facilitate the growth of others in the kingdom and the church,

---

[26]C. H. Dodd, *The Parables of the Kingdom* (London: Nisbet, 1946), 20.
[27]Archibald M. Hunter, *Interpreting the Parables* (Philadelphia: The Westminster Press, 1960), 8.
[28]Archibald M. Hunter, *A Pattern for Life: An Exposition of the Sermon on the Mount* (Philadelphia: The Westminster Press, 1953), 19.
[29]Dodd, *Parables*, 22.
[30]Ibid., 22. Emphasis his.
[31]Hunter, *Pattern*, 19.

then we need to know not only *that* growth occurs but also *how* it occurs so that we may instruct more clearly. The process of reproducing disciples requires that we understand *how* believers grow, develop, and propagate. Jesus informs us through his parables of nature about this process through an appeal to the ecological processes of growth in nature. Seeds planted in the soil, flowers grown in a field, and vines that spread in a vineyard all grow and thrive to full maturity and harvest because they develop and produce while embedded in a natural ecosystem. In this interconnected and highly interactive ecology, individual plants, flowers, seeds, and grapes mature mutually toward their ultimate destinies.

Proposition 5: Christians have a spiritual connection to Christ and other Christians, which forms a spiritual ecology. When Raymond Lindeman conducted his experiments on lake ecosystems, he discovered "the living and nonliving parts of the lake . . . were intimately interconnected."[32] We find it compelling that the apostle Paul goes to great lengths to highlight this very reality in our relationship to Christ and to other members of the body of Christ. We noted in chapters nine and ten the powerful use of language Paul employed to denote and describe how intimately and closely each believer enjoys a connection to Christ and each other. The Greek preposition *syn* ("connected to" or "together with") conveys the closest possible relationship one person can have with another. The preposition, especially when combined with other Greek verbs and nouns, describes a deep solidarity of one with another. Murray Harris declares, "There are more New Testament compounds formed with *syn* than with any other preposition."[33] We highlight this important preposition with its attendant compounds because it establishes a very important ecological necessity. Connections form the cornerstone of any ecosystem linking together the individual living entities in a unified whole. Through these various types of interconnections, a web of life emerges to sustain each individual organism and the ecosystem as a whole.

We observed in chapters nine and ten that Paul's *syn*-compounds divide almost equally into two complementary categories: vertical (redemptive)

---

[32]Frank Benjamin Golley, *A History of the Ecosystem Concept in Ecology: More Than the Sum of the Parts* (New Haven: Yale University Press, 1993), 50.
[33]Murray J. Harris, *Prepositions and Theology* (Grand Rapids: Zondervan, 2012), 199.

syn-compounds that describe how Christians connect to the person and work of Christ and horizontal (relational) *syn*-compounds that describe how Christians connect and relate to each other. Our observation regarding these two uses of *syn*-compounds in Paul finds support in Dunn, who provides two extensive footnotes that list the occurrences of these two categories.[34] Dunn argues that Paul "uses them both to describe the common privilege, experience, and task of believers and to describe a sharing in Christ's death and life."[35] Although we use different terminology, we are describing the same Pauline phenomenon.

The combination of these *syn*-compound categories paints a portrait of a spiritual ecology intertwined and interconnected through its spiritual source of life: Jesus Christ. Just as the sun is the physical source of life for all natural ecosystems, Jesus, the Son of God, is the spiritual source of life for the spiritual ecosystem that bears his name.

Proposition 6: *The spiritual connections we have with other Christians create opportunities for reciprocal exchanges of spiritual nutrients*. Vital as the above connections are to the spiritual vitality of individual Christians and the church as a whole, what happens *between* these connections ensures our ongoing growth and maturity as Christians. Just as interconnections are necessary between the various components of a natural ecosystem, so also are the interactions that form the nutrient pathways governing sustainability. An early pioneer in the study of ecology, Jan Smuts, described the ecosystem as a place where "the whole and the parts therefore reciprocally influence and determine each other."[36] In a similar vein, Paul appeals to the human body to illustrate the interactive relationship of one member to another in the body of Christ (1 Cor 12; Eph 4). John Stoddard, commenting on the organismic quality of the body of Christ, notes that this means therefore that "the members are subject to reciprocal influence."[37] Such is the nature of ecological interactions whether natural or spiritual.

In chapter eleven, we explored Paul's use of another key term in his theology: the frequent use of the reciprocal pronoun *allēlōn*. The term is usually

---

[34]James D. G. Dunn, *The Theology of Paul the Apostle* (Grand Rapids: Eerdmans, 1998), 402-3.
[35]Ibid., 402.
[36]Jan Christiaan Smuts, *Holism and Evolution* (Gouldsboro, ME: Gestalt Journal Press, 2013), 86.
[37]John L. Stoddard, *The Theology of Saint Paul* (Westminster, MD: The Newman Bookshop, 1958), 1:301.

translated "one another" or "each other" and appears frequently in those places in Paul's letters where he wished to emphasize reciprocal relationships between believers, especially as it pertained to the Jew/Gentile relationship. He would often pair *allēlōn* with various *syn*-compounds, thereby emphasizing the necessity for both spiritual connections and interactions to achieve full maturity in Christ (Rom 1:12; 12:16; 1 Cor 12:24-26; Eph 4:2-3; Phil 2:2-3; Col 3:13-14). The *syn*-compound serves as the indicative concerning the reality of the connective relationship as it pertains to either the believer's connection to Christ or to other Christians. The *allēlōn* reciprocal pronoun, along with the accompanying verb, serves as the imperative as believers operationalize the reality of the *syn*-compound connection in their relationship to Christ or other Christians.

Our spiritual connections are a necessary but insufficient component of the spiritual formation process. These *allēlōn* imperatives only come into play in what happens between human connections, and the verbs accompanying each imperative convey how we are to govern our attitudes, actions, and relationships within the body of Christ. When we remain "in Christ," he imparts spiritual nutrients to us through the Spirit, and we in turn share those spiritual nutrients with other believers through reciprocal interactions, exchanges, fellowship, and worship experiences empowered by the same Spirit. For example, Paul writes to the believers in the Philippian church, "I rejoice and share my joy with you all. You too, *I urge you*, rejoice in the same way and share your joy with me" (Phil 2:17-18). Paul describes a mutual or reciprocal exchange of the spiritual nutrient "joy," which we know from Galatians 5:22 is a fruit of the Spirit. Paul has a joy he has received from the Spirit, the Philippian believers have a joy they have received from the same Spirit, and they now share their unique experiences of the joy of the Spirit with each other. In another instance, Paul opens his letter to the Colossians by reminding them that he and Timothy are "praying always for you" and closes his letter by requesting that they in turn keep "praying at the same time for us as well" (Col 1:3; 4:3). Paul not only enjoins the practice of reciprocal exchange of spiritual resources to his churches but also practices it in his own relationship with those churches. He assures the church in Rome that he longs to meet them in person so that they may be "encouraged together . . . each of us by the other's faith, both yours and mine" (Rom 1:12).

Paul is concerned with bringing about whole-person transformation so that everyone may be presented "fully mature in Christ" (Col 1:28 NIV). To that end, he not only encourages the reciprocal exchange of spiritual resources but the exchange of physical and material resources as well. In his second letter to the Corinthians, Paul addresses the relationship between the church at Corinth and the churches of Macedonia, using the language of mutual sharing: "At this present time your abundance *being a supply* for their need, so that their abundance also may become *a supply* for your need, that there may be equality" (2 Cor 8:14). In Galatians, Paul encourages "the one who is taught the word" to also "share all good things with the one who teaches *him*" (Gal 6:6). In his closing remarks to the church at Philippi, Paul commends them for their exceptional service to him by noting that "no church shared with me in the matter of giving and receiving but you alone" (Phil 4:15). Gordon Fee observes that "the language is intended to express both the mutuality and the reciprocity of such 'giving and receiving.'"[38]

Proposition 7: *The spiritual ecology created by Christ through the Spirit is unbounded by time and space, enabling Christians to enjoy the benefits of this reality at any time and in any place, whether in person or online.* Since our spiritual connection to Christ and all Christians comprise the *sanctorum communio,* our ability to connect to Christ and to one another is not limited by time and space. Paul teaches that all Christians enjoy "the fellowship of the Holy Spirit" (2 Cor 13:14; see also Phil 2:1). The English word *fellowship* is an unfortunate translation for the Greek word *koinōnia* that dominates the landscape of Paul's concept of relationships in the church. All forms of the word *koinōnia* derive from one Greek root (*koin*), which refers to something that two or more people share together or have in common.[39] In the two passages above, Paul links *koinōnia* with the Holy Spirit as what all Christians have in common because of their common connection to Christ. Christians share this common experience of the Spirit whether physically absent or present.

---

[38]Gordon Fee, *Paul's Letter to the Philippians* (Grand Rapids: Eerdmans, 1995), 444.
[39]J. Y. Campbell, "*Koinonia* and Its Cognates in the New Testament," *Journal of Biblical Literature* 51 (1932): 352-82; Jeff Kloha, "*Koinonia* and Life Together in the New Testament," *Concordia Journal* 38, no. 1 (2012): 23-32; Andrew T. Lincoln, "Communion: Some Pauline Foundations," *Ecclesiology* 5, no. 2 (2009): 135-60.

While there are many who reject the possibility of fellowship, community, and spiritual growth apart from the physical gathering of believers in a face-to-face encounter, we are not among them. Our study of the ecology of spiritual formation has led us to recognize that God's Spirit can perform his miraculous work of transformation when Christians are gathered and when they are scattered. We come to this conclusion by careful examination of evidence from the physical and social sciences and Scripture. We know, for instance, that humans, apart from any empowerment of the Spirit, can influence each other's beliefs, values, and behaviors, as demonstrated in the study of social networks.[40] One of the earliest and continuing areas of inquiry in social network analysis involves network "tendencies toward reciprocity or mutuality."[41] In particular, social network analysts have charted and measured "the number or proportion of ties in which a specific type of aid is both given and received."[42] The giving and receiving of help, assistance, service and other forms of social exchange occurs frequently, and the net result of these mutual exchanges is mutual influence. As the research on social networks demonstrates,

> influence does not require face-to-face interaction. . . . It encompasses . . . *behavioral contagion* involving the spontaneous pickup or imitation by other[s] . . . as well as *direct influence* in which the actor initiates behavior which has the manifest objective of affecting the behavior of another member of the group.[43]

Notice that the social influence on behavior takes place apart from "face-to-face interaction." Christakis and Fowler's ongoing research on social contagion conclusively reveals that this distant influence includes not only behavior but also attitudes, norms, and beliefs, as we discussed in chapter eight.

If this kind of social influence is taking place among human beings connected in a social network, should we not expect that even more powerful

---

[40]Nicholas A. Christakis and James H. Fowler, *Connected: The Surprising Power of Our Social Networks and How They Shape Our Lives* (New York: Little, Brown, 2009); Stanley Wasserman and Joseph Galaskiewicz, eds., *Advances in Social Network Analysis* (Thousand Oaks, CA: Sage Publications, 1994); Everett M. Rogers, *Diffusion of Innovations*, 5th ed. (New York: Free Press, 2003); Albert-László Barabasi, *Linked* (New York: Plume, 2003).

[41]Stanley Wasserman and Kathrine Faust, *Social Network Analysis: Methods and Applications* (Cambridge: Cambridge University Press, 1994), 13.

[42]Wasserman and Galaskiewicz, *Advances*, 64.

[43]Ibid., 4.

forms of influence are possible among Christians connected in a spiritual network empowered by the Holy Spirit? If humans can influence one another without direct physical contact and face-to-face interaction, should we not expect similar or greater effects among Christians spiritually connected to one another through the "unity of the Spirit?" Paul assumes he can directly influence the members of his distributed congregations through epistolary correspondence, appointed envoys such as Timothy, and prayer. Otherwise, he seems to have wasted much time doing all three. While we certainly do not wish to discount in any way the value and necessity of the gathered community for instruction, worship, fellowship, and service, we would also recognize that our ability to grow in Christ is not limited to these infrequent occasions.

We live in a highly distributed and digitalized world in which our social connections and interactions take place across a broad spectrum of encounters, as we discussed in chapter seven. Heidi Campbell and Stephen Garner have analyzed "the rise of a network society where . . . geography becomes irrelevant as time-space barriers dissolve between people and information."[44] Campbell and Garner help us understand that we are living in an increasingly "multisite reality" as faith is networked like other aspects of our lives in the twenty-first century.[45] As we move into this new network reality, they see that "the online-offline distinction has blurred as the internet has become embedded in everyday realities."[46] This has led to a developing situation where "beliefs and behaviors from offline church traditions are imported into internet environments" and where "online religious practice is intertwined with rather than divorced from traditional religious frameworks."[47] The blended and blurred new reality of offline and online or real and virtual, which includes how we live out and practice our faith, calls for a new way of thinking about how our faith is formed given our present situation. We need a model of spiritual formation that provides a "goodness of fit" with this reality so that we grow in our faith across all sectors of our socio-spiritual ecology. An ecological model of

---

[44]Campbell and Garner, *Networked Theology*, 58.
[45]Ibid., 65.
[46]Ibid., 77.
[47]Ibid., 76.

faith formation "takes seriously the belief that God's involvement with human beings is not limited to the purely physical, everyday world but is also active in the digital locations we create and inhabit."[48] Such a model will also enable Christian educators, pastors, and leaders to avoid limiting the power of their spiritual influence to what transpires within the four walls of their meeting spaces.

Proposition 8: Christians who share a connection to Christ through the Spirit receive an imputed holiness that makes them mutually contagious and provides us with the ability to spread our contagion in online ecologies of learning. In chapter twelve, we argued that social connections to other humans make us socially contagious to each other, enabling mutual influence in terms of attitudes, values, beliefs, and behaviors. We argued further that we see a similar reality emerge from our spiritual connections to one another through Christ that makes us spiritually contagious to each other. We proposed that the revolutionary concept of holiness Jesus revealed in his actions possessed a contagious quality that could infect other people and bring them into fellowship with God.

In 2 Corinthians 7:1, Paul calls for the church in Corinth to engage collectively ("beloved, let us") in the process of "perfecting holiness in the fear of God." Paul reinforces the collective or corporate appeal by his use of the plural participle "perfecting" (*epitelountes*). Hughes translates the present participle as "bringing holiness to completion," thus honoring the force of the root *teleō*, which Louw and Nida translate "to bring an activity to a successful finish."[49] The language of finishing off an activity already started (*epiteleō*) is the same word Paul used in Philippians 1:6: "He who began a good work in you will perfect (*epiteleō*) it until the day of Christ Jesus." As Paul has argued previously in 2 Corinthians 3:1-11, the believers in Corinth are now partakers of and participants in the blessings of the new covenant that had its origin in the Old Testament with Israel and its completion and consummation in Christ with the church ("these promises," 2 Cor 7:1). Those in such a position of privilege have a duty to live according to the

---

[48]Ibid., 96.

[49]R. Kent Hughes, *2 Corinthians: Power in Weakness* (Wheaton, IL: Crossway, 2012), 145; Johannes P. Louw and Eugene A. Nida, *Greek-English Lexicon of the New Testament: Based on Semantic Domains*, 2nd ed. (New York: United Bible Societies, 1989), 658.

ethical demands of this covenant. In 2 Corinthians 7:1, Paul first states the duty negatively, "Let us cleanse ourselves from all defilement of flesh and spirit," and then positively, "perfecting holiness in the fear of God." Redemptive order is always important to note so that we do not erroneously attribute abilities to the flesh that rightfully originate with God through the Spirit. Paul does this in Galatians 3:6-14 with Abraham; he first acknowledged that Abraham "believed God" and only after that did he perform the rite of circumcision (Gen 15; 17). Scott Hafemann captures the force of Paul's theological reasoning when he observes, "Paul did not say, 'Obey the command in order to *become* God's people,' he said, 'Obey the command because you *are* God's people.' That is why all obedience is an expression of the same grace and power that saved his people to begin with."[50]

We have lost the corporate dimension of "perfecting holiness" because we usually think of it in individual categories. Yet as we noted above, the force of the plural forms of address ("beloved" and "us"), as well as the plural forms of "cleanse" and "perfecting," alert us to Paul's perspective that invites the church as a community of faith to participate in this perfecting of holiness. James Howard surveys the Pauline epistles to determine the role of the church community in perfecting holiness and concludes that it "requires active involvement by the community to grow together in progressive sanctification."[51]

In two places, Paul offers intercessory wish prayers for the corporate sanctification of the church in Thessalonica. First, in 1 Thessalonians 3:11-13, Paul prays that God as Father and Jesus as Lord "may establish your hearts without blame in holiness (*hagiōsynē*) before our God and Father at the coming of our Lord Jesus with all His saints." Second, in 1 Thessalonians 5:23-24 Paul prays that the God of peace would "sanctify (*hagiazō*) you entirely; and may your spirit and soul and body be preserved complete, without blame at the coming of our Lord Jesus Christ." We are dealing here with two prayers that possess "a similarity of thought and structure."[52] In regard to similarity of thought, Wiles notes that "in both prayers the apostle desires

---

[50]Scott J. Hafemann, *2 Corinthians*, NIV Application Commentary (Grand Rapids: Zondervan, 2000), 293.

[51]James M. Howard, *Paul, the Community, and Progressive Sanctification* (New York: Peter Lang, 2007), 185.

[52]Gordon P. Wiles, *Paul's Intercessory Prayers* (London: Cambridge University Press, 1974), 64.

for the Thessalonians a perfection of holiness . . . and envisages their whole beings made ready to stand in the presence of God and Christ."[53]

The first thing that strikes us as we read both of these prayers is their corporate focus. This should not be surprising since this is an obvious emphasis throughout the entire epistle (1 Thess 1:1-2; 2:8, 19-20; 3:9-10; 4:9-12). Paul prays for the church as a whole concerning their corporate sanctification. Even Fee, who sees an individual focus in 1 Thessalonians 5:23, says, regarding the language used by Paul, that "the emphasis, after all, lies on *their* being sanctified and on the *entirety* of the human person that needs this sanctifying work of God."[54] We see the corporate or community emphasis specifically in how Paul uses a plural form of address ("brothers and sisters," NIV) and consistently uses plural pronouns both in 1 Thessalonians 3:11-13 and 5:23-24—even though it is impossible to tell in English translations. In his comments on the 1 Thessalonians 5:23-24 passage, Anthony Hoekema notes, "It is most important for us to realize that sanctification is not something we do by ourselves."[55] Although there is an individual responsibility to be a holy person (1 Thess 4:3-4), we strive to manifest this holiness in both our individual lives and in our relationships with other believers and even non-Christians (1 Thess 3:12; 4:12; 5:15). Paul not only prays that the church will achieve holiness but he calls upon them to practice holiness as they "walk and please God," "abstain from sexual immorality," "do not defraud one another," "love one another," "lead a quiet life," "comfort one another," remain "alert and sober," "encourage one another and build up one another," "live in peace with one another," and "seek after that which is good for one another" (1 Thess 5:22-23; see specifically, 1 Thess 4:1, 3, 6, 9, 11, 18; 5:6, 11, 15). The clustering of "one another" (*allēlōn*) admonitions in the latter portion of the epistle signals Paul's understanding that these interpersonal behaviors encourage the manifestation of the holiness they possess as they experience "the will of God, your sanctification." As they reciprocally engage one another in holy living, they collectively reflect God's holiness and mutually spread their holiness to one another.

---

[53]Ibid.

[54]Gordon D. Fee, *The First and Second Letters to the Thessalonians* (Grand Rapids: Eerdmans, 2009), 228-29.

[55]Anthony A. Hoekema, *Saved by Grace* (Grand Rapids: Eerdmans, 1989), 200.

Our ecology-of-faith model provides a comprehensive perspective on the process of spiritual growth that appreciates both its individual as well as corporate expressions. In addition, it reminds us that spiritual growth requires the appropriation of all of the resources God provides. While many excellent studies highlight the role of God's Son, God's Spirit, and God's word, our small contribution reminds us of the value of God's people to achieve the biblical goal of being made complete in Christ (Col 1:28; 2:10; 1 Thess 5:23; Jas 1:4). The insight that we gain from Scripture and from creation reminds us that spiritual growth is always in harmony with the patterns of growth that govern all God has made. He has provided abundant examples in agriculture, horticulture, and viticulture that growth in God's world always requires ecological connections and interactions that vivify and edify God's people, and ultimately glorify God. Our common connection to Christ, which connects us to each other, creates a spiritual ecology in which Christ, through the Spirit, sanctifies and cleanses "by the washing of water with the word, that He might present to Himself the church in all her glory, having no spot or wrinkle or any such thing; but that she would be holy and blameless" (Eph 5:26-27). *Soli Deo Gloria!*

CONCLUSION

Our purpose throughout this presentation of what the Bible teaches regarding how we grow as Christians was to help God's people recognize and appreciate the complexity and beauty of the intricate patterns of growth God has established that govern natural, social, and spiritual ecologies. God's word reveals an ecological motif that shows how natural growth illustrates spiritual growth manifested in the numerous references to trees, flowers, seeds, plants, vineyards, forests, and fields. Jesus taught the multitudes and his disciples using parables that consistently direct attention to various kinds of natural growth to help us understand growth in the church and kingdom. The apostle Paul employed the ecology of the human body and the interconnectedness of its various members to illustrate the inner workings of the body of Christ that makes possible its extensive and intensive growth.

We proposed that this ecological way of thinking about spiritual growth of the church and Christians has significant implications for and applications to our practice of online Christian education. Although some are reluctant or even resistant to acknowledging the possibility of spiritual growth in online communities of faith, our model offers a thoughtful and considered response to those legitimate concerns. It is our hope and prayer that there might continue to be an ongoing conversation about these matters of faith that are so central to our individual institutional missions and the larger mission of God. The experiences we have had as educators—in a number of institutions across a wide spectrum of evangelical persuasion— convinces us that this model has wide application without sacrificing faith traditions or theological positions.

As we look forward into the future of online Christian education, we are more encouraged and hopeful than ever that this particular means of fulfilling the Great Commission will continue to expand numerically and improve qualitatively. We have witnessed how online learning methods and strategies have positively influenced the instructional practices of on-campus faculty. We have been encouraged by faculty members who were initially hesitant about their participation in online education come to champion it and recruit their colleagues on our behalf. We have seen online students who were unsure of their ability to succeed academically in this new approach to learning achieve a level of academic success they could not have imagined. The testimonies and comments received from online students in our courses and degree programs reveal the extent to which their experiences have had a positive impact on their spiritual growth and on their development as whole persons.

There are, in addition to these reasons for optimism about the future of online Christian education, potential areas of concern and caution. It is easy for institutions yet to embrace online learning to see the success of those institutions who have fully embraced it in purely economic categories. The institutional decision to develop online courses and degree programs needs to be made within the stated mission of the institution rather than as an opportunity to improve the bottom line. It is easy to do online education poorly but it is difficult and expensive to do it with excellence and competence. It is our hope and prayer that what we have offered in these pages will contribute to the pursuit of such excellence and the spiritual growth of students, faculty, and staff who constitute ecologies of faith across the landscape of Christian higher education.

BIBLIOGRAPHY

Ackerman, Jennifer. "The Ultimate Social Network." *Scientific American*, June 2012, 36-43.

Adam, Peter. *Hearing God's Word: Exploring Biblical Spirituality.* Downers Grove, IL: InterVarsity Press, 2004.

Ahern, Barnabas M. "Christian's Union with the Body of Christ in Cor, Gal, and Rom." *Catholic Biblical Quarterly* 23, no. 2 (1961): 199-209.

Allee, W. C. *Principles of Animal Ecology.* Philadelphia: W. B. Saunders, 1949.

Alter, Robert, and Frank Kermode. *The Literary Guide to the Bible.* Cambridge: Harvard University Press, 1987.

Arendale, David. "Review of Successful Practices in Teaching and Learning." Unpublished manuscript. University of Missouri, Kansas City, 1998.

Arnold, Bradley. "Re-envisioning the Olympic Games: Paul's Use of Athletic Imagery in Philippians." *Theology* 115, no. 4 (2012): 243-52.

Bailey, Mark L. "The Parable of the Mustard Seed." *Bibliotheca Sacra* 155 (1998): 449-59.

Balswick, Jack O., Pamela Ebstyne King, and Kevin S. Reimer. *The Reciprocating Self: Human Development in Theological Perspective.* Downers Grove, IL: IVP Academic, 2005.

Balz, Horst, and Gerhard Schneider, eds. *Exegetical Dictionary of the New Testament.* Grand Rapids: Eerdmans, 1994.

Banks, Robert. *Paul's Idea of Community: The Early House Churches in Their Historical Setting.* Grand Rapids: Eerdmans, 1980.

Barabási, Albert-László. *Linked: How Everything Is Connected to Everything Else and What It Means for Business, Science, and Everyday Life.* London: Plume, 2003.

Barash, Vladimir. "The Dynamics of Social Contagion." Unpublished PhD dissertation, Graduate School of Cornell University, 2011.

Barna Group. "Cyber Church: Pastors and the Internet." Last updated February 11, 2015. www.barna.com/research/cyber-church-pastors-and-the-internet/.

———. "How Technology Is Changing Millennial Faith." Last updated October 15, 2011. www.barna.org/barna-update/millennials/640-how-technology-is -changing-millennial-faith#.Ul9dGFN21qM.

———. "More Americans Are Seeking Net-Based Faith Experiences." Last updated May 21, 2001. www.barna.com/research/more-americans-are-seeking -net-based-faith-experiences/.

Barth, Karl. *Church Dogmatics*. Peabody, MA: Hendrickson, 2010.

———. "The Holy Spirit and the Gathering of the Christian Community." *Church Dogmatics* IV.I. Peabody, MA: Hendrickson, 2010.

Barth, Markus. *Ephesians: Translation and Commentary on Chapters 4–6*. Anchor Bible. New York: Doubleday, 1974.

Bauckham, Richard. *The Bible and Ecology: Rediscovering the Community of Creation*. Waco, TX: Baylor University Press, 2010.

———. "Reading the Synoptic Gospels Ecologically." In *Ecological Hermeneutics: Biblical, Historical and Theological Perspectives*, ed. David G. Horrell, Cherryl Hunt, Christopher Southgate, and Francesca Stavrakopoulou. London: T&T Clark, 2010.

Beasley-Murray, G. R. *Jesus and the Kingdom of God*. Grand Rapids: Eerdmans, 1986.

Bedell, Ken. "Technology and Theological Education." *Religion Research*. Last updated 1999. www.religion-research.org/Education.htm.

Berger, Jonah. *Contagion: Why Things Catch On*. New York: Simon & Schuster, 2013.

Berger, Peter L. *The Sacred Canopy: Elements of a Sociological Theory of Religion*. New York: Anchor Books, 1967.

Berry, Wendell. *Sex, Economy, Freedom, and Community: Eight Essays*. New York: Pantheon, 1993.

Best, Ernest. *One Body in Christ*. London: SPCK, 1955.

Billings, J. Todd. *Union with Christ: Reframing Theology and Ministry for the Church*. Grand Rapids: Baker Academic, 2011.

Blomberg, Craig L. *Contagious Holiness: Jesus' Meals with Sinners*. New Studies in Biblical Theology. Downers Grove, IL: InterVarsity Press, 2005.

Bock, Darrell L. *Acts*. Baker Exegetical Commentary on the New Testament. Grand Rapids: Baker Academic, 2007.

Bohm, David. *Wholeness and the Implicate Order*. London: Routledge Classics, 2002.

Bonhoeffer, Dietrich. *Sanctorum Communio: A Theological Study of the Sociology of the Church*. Minneapolis: Fortress, 1998.

Breech, James. *Jesus and Postmodernism*. Philadelphia: Fortress, 1989.

Brenner, Athalya. "Naomi and Ruth." *Vetus Testamentum* 33, no. 4 (1983): 385-97.

Bronfenbrenner, Urie. "A Constant Frame of Reference for Sociometric Research." *Sociometry* 6, no. 4 (1943): 363-97.

———. *The Ecology of Human Development: Experiments by Nature and Design.* Cambridge: Harvard University Press, 1979.

———. *Making Human Beings Human.* Thousand Oaks, CA: Sage, 2005.

———. "Toward an Experimental Ecology of Human Development." *American Psychologist*, July 1997, 513-31.

Brown, Raymond E. *An Introduction to the New Testament.* New Haven, CT: Yale University Press, 1997.

Brueggemann, Walter. *Hope Within History.* Atlanta: John Knox Press, 1987.

———. *Praying the Psalms: Engaging Scripture and the Life of the Spirit.* Eugene, OR: Wipf and Stock, 2007.

Burger, Hans. *Being in Christ: A Biblical and Systematic Investigation in a Reformed Perspective.* Eugene, OR: Wipf & Stock, 2009.

Callenbach, Ernest. *Ecology: A Pocket Guide.* Berkeley: University of California Press, 2008.

Campbell, Constantine R. *Paul and Union with Christ: An Exegetical and Theological Study.* Grand Rapids: Zondervan, 2012.

Campbell, Heidi. *Exploring Religious Community Online: We Are One in the Network.* New York: Lang, 2005.

Campbell, Heidi A., and Stephen Garner. *Networked Theology: Negotiating Faith in Digital Culture.* Grand Rapids: Baker Academic, 2016.

Campbell, J. Y. "*Koinonia* and Its Cognates in the New Testament." *Journal of Biblical Literature* 51 (1932): 352-82.

Capra, Fritjof. *The Web of Life: A New Scientific Understanding of Living Systems.* New York: Anchor Books, 1996.

Christakis, Nicholas A., and James H. Fowler. "The Collective Dynamics of Smoking in a Large Social Network." *The New England Journal of Medicine* 358 (2008): 2249-58.

———. *Connected: The Surprising Power of Our Social Networks and How They Shape Our Lives.* New York: Little, Brown and Company, 2009.

———. "The Spread of Obesity in a Large Social Network over 32 Years." *New England Journal of Medicine* 357 (2007): 370-79.

Collomb, Jean-Daniel. "Christianity to Ecology: John Muir's Walk Through America." *Transtext(e)s Transcultures* 4 (2008): 100-113. http://transtexts.revues.org/254 ; DOI : 10.4000/transtexts.254.

Commoner, Barry. *The Closing Circle: Nature, Man, and Technology.* New York: Alfred A. Knopf, 1971.

Conradie, Ernst M. *Christianity and Ecological Theology.* Stellenbosch, South Africa: Sun Press, 2006.

Conzelmann, Hans. *History of Primitive Christianity.* Nashville: Abingdon, 1973.

Cooper, Christopher. *The Truth About Tesla: The Myth of the Lone Genius in the History of Innovation.* New York: Race Point, 2015.

Cormode, Scott. "Using Computers in Theological Education: Rules of Thumb." *Theological Education* 36, no. 1 (1999): 101-16.

Coutts, Frederick. *The Splendor of Holiness.* London: Salvationist Publishing, 1983.

Cox, Brian, and Jeff Forshaw. *The Quantum Universe.* Boston: Da Capo, 2011.

Cranfield, C. E. B. *A Critical and Exegetical Commentary on the Epistle to the Romans.* Vol. 1. International Critical Commentary. Edinburgh: T&T Clark, 1975.

Crossan, John Dominic. "The Seed Parables of Jesus." *Journal of Biblical Literature* 92, no. 2 (1973): 244-66.

Dahl, N. A. "The Parables of Growth." *Studia Theologica* 5, no. 2 (1951): 132-66.

Dalton, William J. "The Integrity of Philippians." *Biblica* 60, no. 1 (1979): 97-102.

Davidson, Cathy. "Collaborative Learning for the Digital Age." *The Chronicle of Higher Education.* August 26, 2011. http://chronicle.com/article/Collabor ative-Learning-for-the/128789/.

Davis, Ellen F. *Scripture, Culture, and Agriculture: An Agrarian Reading of the Bible.* New York: Cambridge University Press, 2009.

Degenne, Alain, and Michel Forsé. *Introducing Social Networks.* London: Sage, 2006.

Delamarter, Steve, and Dan Brunner. "Theological Education and Hybrid Models of Distance Learning." *Theological Education* 40, no. 2, (2004): 145-64.

de Nooy, Wouter, Andrej Mrvar, and Vladimir Batagelj. *Exploratory Social Network Analysis with Pajek.* Cambridge: Cambridge University Press, 2005.

Dick, Walter, Lou Carey, and James O. Carey. *The Systematic Design of Instruction.* 8th ed. London: Pearson, 2014.

Dodd, C. H. *The Parables of the Kingdom.* London: Nisbet & Company, 1946.

Doty, William G. *Letters in Primitive Christianity.* Philadelphia: Fortress Press, 1973.

Dunn, James D. G. *The Theology of Paul the Apostle.* Grand Rapids: Eerdmans, 1998.

Eisenman, Robert. *The Dead Sea Scrolls and the First Christians.* Shaftesbury, UK: Element Books, 1996.

Elazar, Daniel J. "The Political Theory of Covenant: Biblical Origins and Modern Developments." *Publius* 10, no. 4 (1980): 3-30.

Enns, Paul. *Heaven Revealed.* Chicago: Moody Press, 2011.

Etzioni, Amitai. "On Virtual, Democratic Communities." In *Community in the Digital Age: Philosophy and Practice,* edited by Andrew Feenberg and Darin Barney, 225-38. Lanham, MD: Rowman and Littlefield, 2004.

Fee, Gordon D. *The First and Second Letters to the Thessalonians.* New International Commentary on the New Testament. Grand Rapids: Eerdmans, 2009.

———. *The First Epistle to the Corinthians.* New International Commentary on the New Testament. Grand Rapids: Eerdmans, 1987.

———. *God's Empowering Presence: The Holy Spirit in the Letters of Paul.* Peabody, MA: Hendrickson, 1994.

———. *Paul's Letter to the Philippians.* New International Commentary on the New Testament. Grand Rapids: Eerdmans, 1995.

Feenberg, Andrew, and Maria Bakardjieva. "Consumers or Citizens? The Online Community Debate." In *Community in the Digital Age: Philosophy and Practice,* edited by Andrew Feenberg and Darin Barney, 1-30. Lanham, MD: Rowman and Littlefield, 2004.

Field, D. H. "Conceive, Apprehend." In *The New International Dictionary of New Testament Theology,* edited by Colin Brown. Vol. 1. Grand Rapids: Zondervan, 1979.

Fields, Weston. *Sodom and Gomorrah: History and Motif in Biblical Narrative.* Sheffield: Sheffield Academic Press, 2009.

Fong, Bruce William. "Our Mutual Responsibilities Toward One Another as Seen in the Context of *Allelon.*" Master of Theology thesis, Dallas Theological Seminary, May 1978.

Fowler, James H., and Nicholas A. Christakis. "Dynamic Spread of Happiness in a Large Social Network: Longitudinal Analysis over 20 Years in the Framingham Heart Study." *British Medical Journal* 337 (2008): 1-9.

Freedman, William. "The Literary Motif: A Definition and Evaluation." *NOVEL: A Forum on Fiction* 4, no. 2 (Winter 1971): 123-31.

Friederichs, K. "A Definition of Ecology and Some Thoughts About Basic Concepts." *Ecology* 39, no. 1. (1958): 154-59.

Fry, Richard. "Millennials Overtake Baby Boomers as America's Largest Generation." *Pew Research Center.* April 25, 2016. www.pewresearch.org/fact -tank/2016/04/25/millennials-overtake-baby-boomers/.

Garland, David E. *Mark.* The NIV Application Commentary. Grand Rapids: Zondervan, 1996.

Garland, Diana R. *Family Ministry: A Comprehensive Guide.* Downers Grove, IL: IVP Academic, 2012.

Gehring, Roger W. *House Church and Mission: The Importance of Household Structures in Early Christianity.* Peabody, MA: Hendrickson, 2004.

Gibson, James J. *The Ecological Approach to Visual Perception.* New York: Psychology Press, 2015.

Glover, Heather. "Using the Internet to Renew Community Ties in Traditional Religious Communities." PhD dissertation, Baylor University, Waco, TX, 2003. Masters Abstracts International.

Glueck, Nelson. *Hesed in the Bible*. Jersey City, NJ: KTAV Publishing House, 1975.

Goldingay, John. *Psalms*. Vol. 2. Baker Commentary on the Old Testament. Grand Rapids: Baker Academic, 2007.

———. *Psalms*. Vol. 3. Baker Commentary on the Old Testament. Grand Rapids: Baker Academic, 2008.

Golley, Frank Benjamin. *A History of the Ecosystem Concept in Ecology: More Than the Sum of the Parts*. New Haven, CT: Yale University Press, 1993.

Gordon, Benjamin D. "On the Sanctity of Mixtures and Branches: Two Halakic Sayings in Romans 11:16-24." *Journal of Biblical Literature* 135, no. 2 (2016): 355-68.

Granovetter, Mark S. "The Strength of Weak Ties." *American Journal of Sociology* 78, no. 6 (1973): 1360-80.

Green, Michael. *Evangelism in the Early Church*. Grand Rapids: Eerdmans, 1970.

Greenwood, Shannon, Andrew Perrin, and Maeve Duggan. "Social Media Update 2016." Pew Research Center, November 11, 2016. www.pewinternet .org/2016/11/11/social-media-update-2016.

Grenz, Stanley J. *Created for Community: Connecting Christian Belief with Christian Living*. 2nd ed. Grand Rapids: Baker Books, 2001.

———. *Theology for the Community of God*. Grand Rapids: Eerdmans, 1994.

Gresham, John. "The Divine Pedagogy as a Model for Online Education." *Teaching Theology and Religion* 9 (2006): 24-28.

Groothuis, Douglas. "Christian Scholarship and the Philosophical Analysis of Cyberspace Technologies." *Journal of the Evangelical Theological Society* 41 (1998): 631-40.

———. *The Soul in Cyberspace*. Grand Rapids: Baker, 1997.

Gundry, Robert H. *Sōma in Biblical Theology: With Emphasis on Pauline Anthropology*. Grand Rapids: Academie Books, 1976.

Hafemann, Scott J. *2 Corinthians*. NIV Application Commentary. Grand Rapids: Zondervan, 2000.

Hansen, G. Walter. *The Letter to the Philippians*. Pillar New Testament Commentary. Grand Rapids: Eerdmans, 2009.

Harris, Murray J. *Prepositions and Theology in the Greek New Testament*. Grand Rapids: Zondervan, 2012.

Harris, R. Laird, Gleason L. Archer, and Bruce K. Waltke. *Theological Wordbook of the Old Testament.* Vol. 2. Chicago: Moody Press, 1980.

Harvey, John D. "The 'With Christ' Motif in Paul's Thought." *Journal of the Evangelical Theological Society* 35, no. 3 (September 1992): 329-40.

Hedges, Brian G. *Christ Formed in You: The Power of the Gospel for Personal Change.* Wapwallopen, PA: Shepherd Press, 2010.

Henning, Brian, and Adam Scarfe. *Beyond Mechanism: Putting Life Back into Biology.* Lanham, MD: Lexington Books, 2013.

Hepper, F. Nigel. *Baker Encyclopedia of Bible Plants: Flowers and Trees, Fruits and Vegetables, Ecology.* Grand Rapids: Baker, 1992.

Hess, Mary. "Attending to Embodiedness in Online, Theologically Focused Learning." Personal website. October 2000. http://meh.religioused.org/web/Home.html.

Hiebert, Theodore. *The Yahwist's Landscape: Nature and Religion in Early Israel.* New York: Oxford University Press, 1996.

Hill, Andrew E. "The Temple of Asclepius: An Alternative Source for Paul's Body Theology?" *Journal of Biblical Literature* 99, no. 3 (1980): 437-39.

Hillel, Daniel. *The Natural History of the Bible: An Environmental Exploration of the Hebrew Scriptures.* New York: Columbia University Press, 2006.

Hirsch, Alan. *The Forgotten Ways: Reactivating the Missional Church.* Grand Rapids: Brazos, 2006.

Hoch, Carl B. "The Significance of the *Syn*-Compounds for Jew-Gentile Relationships in the Body of Christ." *Journal of the Evangelical Theological Society* 25, no. 2 (1982): 175-83.

Hoehner, Harold W. *Ephesians: An Exegetical Commentary.* Grand Rapids: Baker Academic, 2002.

Hoekema, Anthony A. *Saved by Grace.* Grand Rapids: Eerdmans, 1989.

Hooker, Morna D. "A Partner in the Gospel: Paul's Understanding of His Ministry." *Theology and Ethics in Paul and His Interpreters: Essays in Honor of Victor Paul Furnish,* edited by Eugene H. Lovering and Jerry L. Sumney. Nashville: Abingdon, 1996.

Horrell, David G. *Solidarity and Difference: A Contemporary Reading of Paul's Ethics.* London: T&T Clark, 2005.

Horrigan, John. "Online Communities: Networks That Nurture Long-Distance Relationships and Local Ties." Pew Internet and American Life Project. Last updated October 31, 2001. www.pewinternet.org/pdfs/PIP_Communities_Report.pdf.

Horton, Dennis. *Death and Resurrection: The Shape and Function of a Literary Motif in the Book of Acts.* Eugene, OR: Wipf and Stock, 2009.

Howard, James M. *Paul, the Community, and Progressive Sanctification: An Exploration into Community-Based Transformation Within Pauline Theology.* New York: Peter Lang, 2007.

Hughes, J. Donald. *Ecology in Ancient Civilizations.* Santa Fe: University of New Mexico Press, 1975.

Hughes, R. Kent. *2 Corinthians: Power in Weakness.* Wheaton, IL: Crossway, 2012.

Hunter, Archibald M. *Interpreting the Parables.* Philadelphia: Westminster Press, 1960.

———. *A Pattern for Life: An Exposition of the Sermon on the Mount.* Philadelphia: Westminster Press, 1953.

Illich, Ivan. *Deschooling Society.* New York: Harper and Row, 1970.

Irenaeus. *Saint Irenaeus of Lyons: Against Heresies.* South Bend, IN: Exfontibus Company, 2015.

Jeremias, Joachim. *The Parables of Jesus.* New York: Charles Scribner and Sons, 1972.

Jewell, John P. *Wired for Ministry: How the Internet, Visual Media, and Other New Technologies Can Serve Your Church.* Grand Rapids: Brazos, 2004.

Jewett, Robert. *Paul's Anthropological Terms: A Study of Their Use in Conflict Settings.* Leiden: E. J. Brill, 1971.

Johnson, Aubrey R. *Sacral Kingship in Ancient Israel.* Cardiff: University of Wales Press, 1967.

———. *The Vitality of the Individual in the Thought of Ancient Israel.* Cardiff: University of Wales Press, 1949.

Johnson, Leea. "Paul's Epistolary Presence in Corinth: A New Look at Robert W. Funk's Apostolic Parousia." *Catholic Biblical Quarterly* 68, no. 3 (2006): 481-501.

Jones, Peter R. "The Seed Parables in Mark." *Review and Expositor* 75 (1978): 519-38.

Kaku, Michio. *Physics of the Future: How Science Will Shape Human Destiny and Our Daily Lives by the Year 2100.* New York: Random House, 2011.

Kasemann, Ernst. *Essays on New Testament Themes.* Philadelphia: Fortress, 1964.

———. *Perspectives on Paul.* Philadelphia: Fortress, 1971.

Kear, Karen. *Online and Social Networking Communities: A Best Practice Guide for Educators.* New York: Routledge, 2011.

Keck, Leander. *Paul and His Letters.* Philadelphia: Fortress, 1979.

Kemp, Steve. "Learning Communities in Distance Education." Paper presented at the conference of the Association of Christian Continuing Education of Schools and Seminaries, Seal Beach, CA, 2002.

———. "Social Presence in Online Learning." *Best Practices of Online Education: A Guide for Christian Higher Education*, edited by Mark A. Maddix et al., 41-54. Charlotte: Information Age Publishing, 2012.

Kirsch, J. P. *The Doctrine of the Communion of Saints in the Ancient Church*. Middletown, DE: Christ the King Library, 1911.

Kistemaker, Simon J. *The Parables of Jesus*. Grand Rapids: Baker, 1980.

Klingbeil, Martin G. "Creation in the Prophetic Literature of the Old Testament: An Intertextual Approach." *Journal of the Adventist Theological Society* 12, nos. 1-2 (2009): 19-54.

Kloha, Jeff. "*Koinonia* and Life Together in the New Testament." *Concordia Journal* 38, no. 1 (2012): 23-32.

Kramer, Adam D. I., Jamie E. Guillory, and Jeffrey T. Hancock. "Experimental Evidence of Massive Scale Emotional Contagion Through Social Networks." *Proceedings of the National Academy of Sciences* 111, no. 29 (2014): 8788-90.

Krämer, H. "*Allēlōn*." In *Exegetical Dictionary of the New Testament*, edited by Horst Balz and Gerhard Schneider. Vol. 1. Grand Rapids: Eerdmans, 1999.

Kraus, C. Norman. *The Authentic Witness*. Grand Rapids: Eerdmans, 1979.

Kurian, George Thomas. "Communion of the Saints." *Encyclopedia of Christian Education*, edited by George Thomas Kurian and Mark A. Lamport. Lanham, MD: Roman and Littlefield, 2015.

Kuyper, Lester J. "Grace and Truth." *Reformed Review* 16, no. 1 (1962): 1-16.

Larsen, Elena. "Cyberfaith: How Americans Pursue Religion Online." Pew Internet and American Life Project. December 23, 2001. www.pewinternet.org/pdfs/PIP_CyberFaith_Report.pdf.

Laszlo, Ervin. *The Interconnected Universe: The Conceptual Foundations of Transdisciplinary Unified Theory*. Singapore: World Scientific Publishing, 1995.

Lee, Moira. "Experiencing Shared Inquiry Through the Process of Collaborative Learning." *Teaching Theology and Religion* 3, no. 2 (2000): 108-16.

Lehman, Rosemary, and Simone Conceicao. *Creating a Sense of Presence: How to Be There for Online Students*. San Francisco: John Wiley & Sons, 2010.

Lerner, Richard. *Concepts and Theories of Human Development*. Mahwah, NJ: Lawrence Erlbaum Associates, 2002.

Leroy, H. "*Auxanō, auxō*." In *Exegetical Dictionary of the New Testament*, edited by Horst Balz and Gerhard Schneider. Vol. 1. Grand Rapids: Eerdmans, 1999.

Letham, Robert. *Union with Christ in Scripture, History, and Theology*. Philadelphia: P&R, 2011.

Lightfoot, J. B. *St. Paul's Epistle to the Philippians*. Grand Rapids: Zondervan, 1973.

Lincoln, Andrew T. "Communion: Some Pauline Foundations." *Ecclesiology* 5 (2009): 135-60.

Lock, Jennifer. "Laying the Groundwork for the Development of Learning Communities Within Online Courses." *Quarterly Review of Distance Education* 3 (2002): 395-408.

Lohfink, Gerhard. *Jesus and Community*. Philadelphia: Fortress, 1984.

Louw, Johannes P., and Eugene A. Nida. *Greek-English Lexicon of the New Testament Based on Semantic Domains*. New York: United Bible Societies, 1989.

Lowe, Mary E. "Assessing the Impact of Online Courses on the Spiritual Formation of Adult Students." Unpublished EdD dissertation, Nova Southeastern University, 2007.

Luo, Yiqi, Ensheng Weng, and Yuanhe Yang. "Ecosystem Ecology." In *Encyclopedia of Theoretical Ecology*, edited by Alan Hastings and Louis J. Gross. Berkeley: University of California Press, 2012.

Lynch, James J. *A Cry Unheard: New Insights into the Medical Consequences of Loneliness*. Baltimore, MD: Bancroft Press, 2000.

Lytle, Julie. *Faith Formation 4.0: Introducing an Ecology of Faith in a Digital Age*. New York: Morehouse Publishing, 2013.

Macaskill, Grant. *Union with Christ in the New Testament*. Oxford: Oxford University Press, 2013.

Maddix, Mark. *Spiritual Formation: A Wesleyan Paradigm*. Kansas City, MO: Beacon Hill Press, 2011.

Magnusson, David, and Vernon Allen. *Human Development: An Interactional Perspective*. New York: Harcourt Brace Jovanovich, 1983.

Malina, Bruce J. *The New Testament World: Insights from Cultural Anthropology*. Louisville, KY: Westminster John Knox, 2001.

Marchal, Joseph A. *Hierarchy, Unity, and Imitation: A Feminist Rhetorical Analysis of Power Dynamics in Paul's Letter to the Philippians*. Atlanta: Society of Biblical Literature, 2006.

Margalef, R. "Networks in Ecology." In *Theoretical Studies of Ecosystems: The Network Perspective*, edited by M. Higashi and T. P. Burns. New York: Cambridge University Press, 1991.

Marshall, Peter. *Enmity in Corinth: Social Conventions in Paul's Relations with the Corinthians*. Tubingen: Mohr Siebeck Verlag, 1987.

Mattis, Jacqueline S., Muninder K. Ahluwalia, Sheri-Ann E. Cowie, and Aria M. Kirkland-Harris. "Ethnicity, Culture, and Spiritual Development." In *The Handbook of Spiritual Development in Childhood and Adolescence*, edited by

Eugene C. Roehlkepartain, Pamela Ebstyne King, Linda Wagener, and Peter L. Benson. San Francisco: Sage, 2006.

McGourty, G. T., J. Ohmart, and D. Chaney. *Organic Winegrowing Manual*. Richmond, CA: The University of California Division of Agriculture and Natural Resources, 2011.

McGrath, Brendan. "The Doctrine of Christian Solidarity in the Epistles of Saint Paul." Unpublished doctor of sacred theology degree, University of Ottawa, 1952.

———. "'Syn' Words in Saint Paul." *Catholic Biblical Quarterly* 14 (July 1952): 219-26.

McKenzie, Roderick D. *On Human Ecology*. Chicago: University of Chicago Press, 1968.

Meadows, Philip. "Mission and Discipleship in a Digital Culture." *Mission Studies* 29 (2012): 163-82.

Meyer, David. "Creating a More Viable Digital Ecosystem for Faith-Based Humanitarian Groups Working in Kenya to Foster Better Support for Potential Benefactors for Tangible Needs." Master's thesis, Savannah College of Art and Design, 2014.

Milgrom, Jacob. "Profane Slaughter and a Formulaic Key to the Composition of Deuteronomy." *Hebrew Union College Annual* 47 (1976): 1-17.

Milne, Esther. *Letters, Postcards, Email: Technologies of Presence*. New York: Routledge, 2010.

Mitchell, Margaret M. "New Testament Envoys in the Context of Greco-Roman Diplomatic and Epistolary Conventions: The Example of Timothy and Titus." *Journal of Biblical Literature* 111, no. 4 (1992): 641-62.

Moll, Rob. "Blogger Predicts Revival via Web: Is the Next Great Awakening Happening on the Internet?" *Christianity Today*. 2004. www.christianity today.com/pastors/2004/fall/4.13.html.

Moller, Leslie, Gustavo E. Prestera, Douglas Harvey, Margaret Downs-Keller, and Jo-Ann McCausland. "Creating an Organic Knowledge-Building Environment Within an Asynchronous Distributed Learning Context." *Quarterly Review of Distance Education* 3 (2002): 47-58.

Montague, George T. *Growth in Christ: A Study in Paul's Theology of Progress*. Kirkwood, MO: Maryhurst Press, 1961.

Moo, Douglas J. *The Epistle to the Romans*. New International Commentary on the New Testament. Grand Rapids: Eerdmans, 1996.

Moon, J. Robin, M. Maria Glymour, Anusha M. Vable, Sze Y. Liu, and S. V. Subramanian. "Short- and Long-Term Associations Between Widowhood and Mortality in the United States: Longitudinal Analyses." *Journal of Public Health* 36, no. 3 (2014): 382-89.

Moore, Michael G. "Three Types of Interaction." *American Journal of Distance Education* 3, no. 2 (1989): 1-4.

Moreno, Jacob. *Who Shall Survive? A New Approach to the Problem of Human Interrelationships.* Washington: Nervous and Mental Diseases Publishing Company, 1934.

Morris, Norval, and David J. Rothman. *The Oxford History of the Prison: The Practice of Punishment in Western Society.* Oxford: Oxford University Press, 1995.

Motyer, J. Alec. *The Prophecy of Isaiah: An Introduction and Commentary.* Downers Grove, IL: InterVarsity Press, 1993.

Mowinckel, S. *He That Cometh: The Messiah Concept in the Old Testament and Later Judaism.* Uppsala: Almqvist & Wiksells, 1943.

Needham, Phil. "Integrating Holiness and Community: The Task of an Evolving Salvation Army." *Word and Deed* 3, no. 1 (2000): 5-20.

Neumayer, Christina, and Celina Raffl. "Facebook for Global Protest: The Potential and Limits of Social Software for Grassroots Activism." Paper presented at the Community Informatics Conference: ICTs for Social Inclusion, 2008.

Oberlinner, L. "*Dólos.*" In *Exegetical Dictionary of the New Testament*, edited by Horst Balz and Gerhard Schneider. Grand Rapids: Eerdmans, 1999.

Oden, Thomas C. *Life in the Spirit.* Peabody, MA: Prince Press, 2001.

Ogereau, Julien M. "A Survey of Koinonia and Its Cognates in Documentary Sources." *Novum Testamentum* 57 (2015): 275-94.

Oldenburg, Ray. *The Great Good Place: Cafés, Coffee Shops, Bookstores, Bars, Hair Salons, and Other Hangouts at the Heart of a Community.* New York: Marlowe, 1989.

Page, Charles R., II. *Jesus and the Land.* Nashville: Abingdon, 1995.

Palka, John. "Defining a Theological Education Community." *International Review of Research in Open and Distance Learning.* November 2004. www.irrodl.org/index.php/irrodl/article/view/197/279.

Palloff, Rena, and Keith Pratt. *Building Learning Communities in Cyberspace.* San Francisco: Jossey-Bass, 1999.

———. *Lessons from the Cyberspace Classroom.* San Francisco: Jossey-Bass, 2001.

Pannenberg, Wolfhart. *Anthropology in Theological Perspective.* Philadelphia: Westminster Press, 1985.

Patten, Bernard C. "Network Ecology: Indirect Determination of the Life-Environment Relationship in Ecosystems." In *Theoretical Studies of Ecosystems: The Network Perspective*, edited by M. Higashi and T. P. Burns. New York: Cambridge University Press, 1991.

Patterson, Elizabeth. "The Questions of Distance Education." *Theological Education* 33, no. 1 (1996): 59-74.

Pedersen, Johs. *Israel: Its Life and Culture.* Vols. I-II. London: Geoffrey Cumberlege, 1964.

Perliger, Arie, and Ami Pedahzur. "Social Network Analysis in the Study of Terrorism and Political Violence." *Political Science and Politics* 44, no. 1 (2011): 45-50.

Perriman, Andrew. "The Corporate Christ: Reassessing the Jewish Background." *Tyndale Bulletin* 50, no. 2 (1999): 239-63.

Pew Research Center. "Religion and Electronic Media: One-in-Five Americans Share Their Faith Online." November 6, 2014. www.pewforum.org/2014/11/06/religion-and-electronic-media/.

Pfitzner, Victor C. *Paul and the Agon Motif.* Leiden: E. J. Brill, 1967.

Prensky, Marc. "Digital Natives, Digital Immigrants." *On the Horizon* 9, no. 5 (2001): 1-6.

Putnam, Robert. *Bowling Alone: The Collapse and Revival of American Community.* New York: Simon & Schuster, 2000.

Putnam, Robert, and Lewis Feldstein (with Don Cohen). *Better Together: Restoring the American Community.* New York: Simon & Schuster, 2003.

Raptis, Dimitrios, Jesper Kjeldskof, Mikael B. Skov, and Jeni Paay. "What Is a Digital Ecology? Theoretical Foundations and a Unified Definition." *Australian Journal of Intelligent Information Processing Systems* 13, no. 4 (2014). https://cs.anu.edu.au/ojs/index.php/ajiips/article/view/1259.

Raymond, Jonathan S. "Social Holiness: Journey, Exposures, Encounters." In *The Holiness Manifesto*, edited by Kevin W. Mannoia and Don Thorsen. Grand Rapids: Eerdmans, 2008.

Reid, Marty L. "A Consideration of the Function of Romans 1:8-15 in Light of Greco-Roman Rhetoric." *Journal of the Evangelical Theological Society* 38, no. 2 (1995): 181-91.

Reinhartz, Adele. "On the Meaning of the Pauline Exhortation '*mimētai mou ginesthe*'—Become Imitators of Me." *Studies in Religion* 16, no. 4 (1987): 393-403.

Rheingold, Howard. *The Virtual Community: Homesteading on the Electronic Frontier.* Cambridge, MA: MIT Press, 1993.

Rice, Jesse. *The Church of Facebook: How the Hyper-Connected Are Redefining Community.* Colorado Springs, CO: David C. Cook, 2009.

Richards, E. Randolph, and Brandon J. O'Brien. *Misreading Scripture with Western Eyes.* Downers Grove, IL: InterVarsity Press, 2012.

Ridderbos, Herman. *Paul: An Outline of His Theology.* Grand Rapids: Eerdmans, 1975.

Rizvi, S. J. H., and V. Rizvi, eds. *Allelopathy: Basic and Applied Aspect.* London: Chapman and Hall, 1992.

Robertson, A. T. *A Grammar of the Greek New Testament in the Light of Historical Research.* Nashville: Broadman, 1934.

Robinson, H. Wheeler. *The Christian Doctrine of Man.* Edinburgh: T&T Clark, 1934.

———. *Corporate Personality in Ancient Israel.* Philadelphia: Fortress, 1980.

———. *Religious Ideas of the Old Testament.* London: Duckworth, 1949.

Robinson, John A. T. *The Body: A Study in Pauline Theology.* London: SCM Press, 1952.

Roebuck, Carl. *Corinth: The Asklepieion and Lerna.* Princeton, NJ: The American School of Classical Studies at Athens, 1951.

Rogers, Everett M. *Diffusion of Innovation.* 5th ed. New York: Free Press, 2003.

Rogers, Paul. *Disturbance Ecology and Forest Management: A Review of the Literature.* Washington: United States Department of Agriculture, 1996.

Rogerson, J. "The Hebrew Conception of Corporate Personality: A Re-examination." *Journal of Theological Studies* 21 (1970): 1-16.

Rosenqust, J. Niels, Joanne Murabito, James H. Fowler, and Nicholas Christakis. "The Spread of Alcohol Consumption Behavior in a Large Social Network." *Annals of Internal Medicine* 152 (2011): 426-33.

Ryan, Edward J. "Ecology: A Study of Interdependence." *The Kansas School Naturalist* 11, no. 1 (1964): 3-14.

Ryken, Leland, James C. Wihoit, and Tremper Longman III, eds. *Dictionary of Biblical Imagery.* Downers Grove, IL: IVP Academic, 1998.

Sameroff, Arnold, ed. *The Transactional Model of Development.* Washington, DC: American Psychological Association, 2009.

Sampley, J. Paul. *Pauline Partnership in Christ.* Philadelphia: Fortress, 1980.

———. "Reasoning from the Horizons of Paul's Thought World: A Comparison of Galatians and Philippians." In *Theology and Ethics in Paul and His Interpreters: Essays in Honor of Victor Paul Furnish,* edited by Eugene H. Lovering Jr. and Jerry L. Sumney. Nashville: Abingdon, 1996.

Samra, James G. *Being Conformed to Christ in Community.* New York: T&T Clark, 2006.

Sanders, Boykin. "Imitating Paul: 1 Cor 4:16." *Harvard Theological Review* 54 (1981): 353-63.

Santmire, H. Paul. *The Travail of Nature: The Ambiguous Ecological Promise of Christian Theology*. Philadelphia: Fortress, 1998.

Schaeffer, Francis A. *Pollution and the Death of Man: The Christian View of Ecology*. Wheaton, IL: Tyndale House Publishers, 1970.

Schneck, Daniel J. "What Is This Thing Called 'Me'? The Stationary, Buffered, Isothermal Living Engine." *American Laboratory* 38, no. 10 (2006): 4-10.

Schnelle, Udo. *Apostle Paul: His Life and Theology*. Grand Rapids: Baker Academic, 2003.

Schubert, Phil. "Grasping the Realities of Educating in the Digital Age." *Educause Review*, April 7, 2011. www.educause.edu/ero/article/grasping-realities -educating-digital-age.

Schultze, Quentin J., and Robert H. Woods Jr., eds. *Understanding Evangelical Media: The Changing Face of Christian Communication*. Downers Grove, IL: IVP Academic, 2008.

Schwarz, Astrid, and Kurt Jax, eds. *Ecology Revisited: Reflecting on Concepts, Advancing Science*. Dordrecht: Springer Science, 2011.

Schwarz, Christian. *Paradigm Shift in the Church*. Carol Stream, IL: ChurchSmart Resources, 1999.

Scott, James W. "The Misunderstood Mustard Seed." *Trinity Journal* 36, no. 1 (2015): 25-48.

Shanks, Hershel. *Understanding the Dead Sea Scrolls*. New York: Random House, 1992.

Shults, F. LeRon. "The 'Body of Christ' in Evangelical Theology." *Word and World* 22, no. 2 (2002): 178-85.

Simard, Suzanne W., David A. Perry, Melanie D. Jones, David D. Myrold, Daniel M. Durall, and Randy Molina. "Net Transfer of Carbon Between Ectomycorrhizal Tree Species in the Field." *Nature* 338, no. 7 (1997): 579-82.

Sinclair, A. R. E., and Charles J. Krebs. "Trophic Interactions, Community Organization, and Kluane Ecosystem." *Ecosystem Dynamics of the Boreal Forest: The Kluane Project*, edited by Charles J. Krebs, Stan Boutin, and Rudy Boustra. New York: Oxford University Press, 2001.

Smith, Marc, and Peter Kollock. *Communities in Cyberspace*. London: Routledge, 2001.

Smith, Robert Leo. *Ecology and Field Biology*. Menlo Park, CA: Addison Wesley Longman, 1996.

Smuts, Jan Christian. *Holism and Evolution*. Gouldsboro, ME: The Gestalt Journal Press, 1986.

Snyder, Howard A. *Decoding the Church: Mapping the DNA of Christ's Body.* Grand Rapids: Baker Books, 2002.

——. *Liberating the Church: The Ecology of Church and Kingdom.* Downers Grove, IL: InterVarsity Press, 1983.

——. *Radical Renewal: The Problem of Wineskins Today.* Eugene, OR: Wipf and Stock, 1996.

——. *Salvation Means Creation Healed: The Ecology of Sin and Grace.* Eugene, OR: Cascade, 2011.

Son, Sang-Won. *Corporate Elements in Pauline Anthropology: A Study of Selected Terms, Idioms, and Concepts in the Light of Paul's Usage and Background.* Rome: Editrice Pontificio Istituto Biblico, 2001.

Song, Felicia Wu. *Virtual Communities: Bowling Alone, Online Together.* New York: Peter Lang, 2009.

Stadd, Allison. "79% of People 18-44 Have Their Smartphones with Them 22 Hours a Day." *Adweek*, April 2, 2013. www.adweek.com/socialtimes/smart-phones/480485.

Stark, Rodney. *The Rise of Christianity.* New York: Harper One, 1997.

Stavredes, Tina. *Effective Online Teaching: Foundations and Strategies for Student Success.* San Francisco: Jossey-Bass, 2011.

Stettler, Hanna. "Sanctification in the Jesus Tradition." *Biblica* 85 (2004): 153-78.

Stoddard, John L. *The Theology of Saint Paul.* Vol. 1. Westminster, MD: Newman Bookshop, 1958.

Tan, Kim-Huat. "Mustard." In *The New Interpreters Dictionary of the Bible*, edited by Katharine Doob Sakenfeld. Vol. 4. Nashville: Abingdon, 2009.

Tannehill, Robert C. *Dying and Rising with Christ: A Study in Pauline Theology.* Eugene, OR: Wipf and Stock, 2006.

Tansley, Alfred George. "The Use and Abuse of Vegetational Concepts and Terms." *Ecology* 16, no. 3 (1935): 284-307.

Taylor, Vincent. *Forgiveness and Reconciliation.* 2nd ed. London: St. Martin's Press, 1952.

Taylor, Walter P. "Significance of the Biotic Community in Ecological Studies." *The Quarterly Review of Biology* 10, no. 3 (1935): 291-307.

Thayer, Joseph Henry. *Greek-English Lexicon of the New Testament.* New York: American Book Company, 1889.

Thielman, Frank. *Philippians.* NIV Application Commentary. Grand Rapids: Zondervan, 1995.

Thompson, Clive. "Are Your Friends Making You Fat?" *New York Times Magazine*, September 10, 2009. www.nytimes.com/2009/09/13/magazine/13contagion -t.html.

Thompson, James W. *Pastoral Ministry According to Paul: A Biblical Vision.* Grand Rapids: Baker Academic, 2006.

Tillich, Paul. *Biblical Religion and the Search for Ultimate Reality.* Chicago: University of Chicago Press, 1955.

Travers, Jeffrey, and Stanley Milgram. "An Experimental Study of the Small World Problem." *Sociometry* 32, no. 4 (1969): 425-43.

Utz, Rebecca L., Deborah Carr, Randolph Nesse, and Camille B. Wortman. "The Effect of Widowhood on Older Adults' Social Participation: An Evaluation of Activity, Disengagement, and Continuity Theories." *Gerontologist* 42, no. 4 (2002): 522-33.

van Aarde, Timothy A. "The Use of *Oikonomia* for Missions in Ephesians." *Verbum et Ecclesia* 37, no. 1 (2016): a1489. http://dx.doi.org/10.4102/ve.v37i1.1489.

VanderKam, James, and Peter Flint. *The Meaning of the Dead Sea Scrolls: Their Significance for Understanding the Bible, Judaism, Jesus, and Christianity.* New York: Harper One, 2002.

Volf, Miroslav. *Exclusion and Embrace: A Theological Exploration of Identity, Otherness, and Reconciliation.* Nashville: Abingdon, 1998.

Von Rad, Gerhard. *Genesis.* Rev. ed. Old Testament Library. Philadelphia: Westminster Press, 1972.

Vygotsky, Lev. *Thought and Language.* Cambridge, MA: MIT Press, 1986.

Walsh, Carey Ellen. *The Fruit of the Vine: Viticulture in Ancient Israel.* Winona Lake, IN: Eisenbrauns, 2000.

Waltke, Bruce K. *Genesis: A Commentary.* Grand Rapids: Zondervan, 2001.

Ward, Richard F. "Pauline Voice and Presence as Strategic Communication." *Semeia* 64 (1994): 95-107.

Ward, Ted W., and Samuel F. Rowen. "The Significance of the Extension Seminary." *Evangelical Missions Quarterly* 9, no. 2 (1972): 17-27.

Wasserman, Stanley, and Joseph Galaskiewicz, eds. *Advances in Social Network Analysis.* Thousand Oaks, CA: Sage, 1994.

Wasserman, Stanley, and Kathrine Faust. *Social Network Analysis: Methods and Applications.* Cambridge: Cambridge University Press, 1994.

Wayne, David. "Tim Bednar's Paper: We Know More Than Our Pastors." *Jolly Blogger*, July 1, 2005. http://jollyblogger.typepad.com/jollyblogger/2005/07/tim_bednars_pap.html#more.

White, Ken. "Face to Face in the Online Classroom." In *The Online Teaching Guide*, edited by Bob Weight and Ken White. Needham Heights, MA: Allyn and Bacon, 2000.

Whitehead, Alfred. *Science and the Modern World*. Cambridge: Cambridge University Press, 1953.

Whitney, Donald S. *Spiritual Disciplines for the Christian Life*. Colorado Springs, CO: NavPress, 1994.

Wiles, Gordon P. *Paul's Intercessory Prayers: The Significance of the Intercessory Prayer Passages in the Letters of Paul*. Cambridge: Cambridge University Press, 1974.

Wilhoit, James C., and John M. Dettoni, eds. *Nurture That Is Christian: Developmental Perspectives on Christian Education*. Wheaton, IL: BridgePoint, 1995.

Wilkinson, Jennifer. "Mark and His Gentile Audience: A Traditio-Historical and Socio-Cultural Investigation of Mk 4.35-9.29 and Its Interface with Gentile Polytheism in the Roman Near East." Unpublished PhD dissertation, Durham University, 2012.

Willis, Avery T., Jr., and Henry T. Blackaby. *On Mission with God: Living God's Purpose for His Glory*. Nashville: B&H, 2002.

Wimberley, Edward T. *Nested Ecology: The Place of Humans in the Ecological Hierarchy*. Baltimore: Johns Hopkins University Press, 2009.

Worster, Donald. *Nature's Economy: A History of Ecological Ideas*. 2nd ed. New York: Cambridge University Press, 1994.

Wright, N. T. Foreword to Marcus J. Borg, *Conflict, Holiness and Politics in the Teachings of Jesus*. Harrisburg, PA: Trinity International Press, 1998.

———. *Paul and the Faithfulness of God*. Parts III and IV. Christian Origins and the Question of God. Vol. 4. Minneapolis: Fortress, 2013.

Yamauchi, Edwin M. "Ancient Ecologies and the Biblical Perspective." *Journal of the American Scientific Affiliation* 32, no. 4 (1980): 193-202.

Yorke, Gosnelll L. O. R. *The Church as the Body of Christ in the Pauline Corpus*. Lanham: University of America Press, 1991.

Zerbe, Gordon. "Ecology According to the New Testament." *Direction* 21, no. 2 (1992): 15-26.

AUTHOR INDEX

Ackerman, Jennifer, 53
Adam, Peter, 196, 197
Ahern, Barnabas M., 141
Allee, W. C., 14
Allen, Vernon, 61, 69
Alter, Robert, 39
Archer, Gleason L., 34
Arendale, David, 101
Arnold, Bradley, 164
Bailey, Mark L., 47
Bakardjieva, Maria, 70
Balswick, Jack O., 182, 186
Banks, Robert, 74
Barabási, Albert-László, 122, 124, 189
Barash, Vladimir, 125
Barth, Karl, 61, 73, 201
Barth, Markus, 197, 198
Batagelj, Vladimir, 125
Bauckham, Richard, 19, 28, 30, 39, 40
Beasley-Murray, G. R., 40
Bedell, Ken, 92
Berger, Jonah, 189
Berger, Peter L., 15
Berry, Wendell, 24
Best, Ernest, 52, 59, 138, 141, 142, 146, 209
Billings, J. Todd, 150
Blackaby, Henry T., 208
Blomberg, Craig L., 188, 191
Bock, Darrell L., 194
Bohm, David, 105
Bonhoeffer, Dietrich, 174, 175, 176, 179, 186
Breech, James, 38
Brenner, Athalya, 178
Bronfenbrenner, Urie, 7, 16, 17, 61, 69, 108, 109, 122, 173, 174, 179, 185, 186
Brown, Raymond E., 151

Brueggemann, Walter, 36
Brunner, Dan, 91, 92
Burger, Hans, 140
Callenbach, Ernest, 126, 212, 214
Campbell, Constantine R., 150
Campbell, Heidi, 3, 79-81, 83, 123, 124, 207, 221
Campbell, J. Y., 157, 219
Capra, Fritjof, 14, 56, 105, 213
Carey, James O., 89, 90
Carey, Lou, 89, 90
Carr, Deborah, 133
Chaney, D., 30
Christakis, Nicholas A., 15, 16, 76, 122, 127, 128, 129, 130, 189, 220
Cohen, Don, 68, 76, 81
Collomb, Jean-Daniel, 14, 24
Commoner, Barry, 138
Conceicao, Simone, 73
Conradie, Ernst M., 25
Conzelmann, Hans, 194
Cooper, Christopher, 208
Cormode, Scott, 91
Coutts, Frederick, 200
Cox, Brian, 3, 11
Cranfield, C. E. B., 139
Crossan, John Dominic, 46
Dahl, N. A., 40, 41
Dalton, William J., 152
Davidson, Cathy, 96, 97
Davis, Ellen F., 27
de Nooy, Wouter, 125
Degenne, Alain, 125
Delamarter, Steve, 91, 92
Dick, Walter, 89, 90
Dodd, C. H., 41, 214, 215
Doty, William G., 110
Duggan, Maeve, 94

Dunn, James D. G., 52, 55, 60, 139, 141, 143, 151, 156, 209, 217
Eisenman, Robert, 189
Elazar, Daniel J., 176, 177
Enns, Paul, 145
Etzioni, Amitai, 69, 70
Faust, Kathrine, 220
Fee, Gordon D., 55, 56, 69, 151, 152, 153, 156, 157, 160, 161, 162, 163, 165, 167, 181, 219, 224
Feenberg, Andrew, 70
Feldstein, Lewis, 68, 76, 81
Field, D. H., 163
Fields, Weston, 26
Flint, Peter, 189
Fong, Bruce William, 180
Forsé, Michel, 125
Forshaw, Jeff, 3, 11
Fowler, James H., 15, 16, 76, 122, 127, 128, 129, 130, 189, 220
Freedman, William, 26
Friederichs, K., 14
Fry, Richard, 114
Garland, David E., 42
Garland, Diana R., 109, 110
Garner, Stephen, 3, 123, 124, 207, 221
Gehring, Roger W., 57
Gibson, James J., 211
Glover, Heather, 71
Glueck, Nelson, 177
Goldingay, John, 28, 32
Golley, Frank Benjamin, 216
Gordon, Benjamin D., 195
Granovetter, Mark S., 125
Green, Michael, 192
Greenwood, Shannon, 94
Grenz, Stanley J., 54, 78, 182

Gresham, John, 82, 83, 112
Groothuis, Douglas, 72, 73, 82
Guillory, Jamie E., 128
Gundry, Robert H., 141
Hafemann, Scott J., 196, 223
Hancock, Jeffrey T., 128
Hansen, G. Walter, 152, 161, 165
Harris, Murray J., 45, 138, 142, 216
Harris, R. Laird, 34
Harvey, John D., 141, 142
Hedges, Brian G., 170
Henning, Brian, 69
Hepper, F. Nigel, 45, 47
Hess, Mary, 79, 107
Hiebert, Theodore, 18, 19
Hill, Andrew E., 54
Hillel, Daniel, 27, 29, 45, 190
Hirsch, Alan, 194, 210
Hoch, Carl B., 183,
Hoehner, Harold W., 58, 59, 145, 197, 198
Hoekema, Anthony A., 200, 224
Hooker, Morna D., 147, 154, 155, 161
Horrell, David G., 39, 40, 164
Horrigan, John, 82, 83, 117
Horton, Dennis, 26
Howard, James M., 131, 186, 196, 201, 223
Hughes, J. Donald, 14
Hughes, R. Kent, 222
Hunter, Archibald M., 41, 56, 186, 187, 215
Illich, Ivan, 85
Irenaeus, 200
Jeremias, Joachim, 38
Jewell, John P., 106, 110
Jewett, Robert, 160
Johnson, Aubrey R., 209
Johnson, Leea, 111
Jones, Peter R., 44
Kaku, Michio, 118
Kasemann, Ernst, 53, 54, 169
Kear, Karen, 94, 95
Keck, Leander, 110
Kemp, Steve, 102
Kermode, Frank, 39
King, Pamela Ebstyne, 182, 186
Kirsch, J. P., 73

Kistemaker, Simon J., 38, 39
Klingbeil, Martin G., 30
Kloha, Jeff, 219
Kollock, Peter, 75, 84
Kramer, Adam D. I., 128
Krämer, H., 179, 180
Kraus, C. Norman, 208
Krebs, Charles J., 172
Kurian, George Thomas, 74
Kuyper, Lester J., 177
Larsen, Elena, 71, 117
Laszlo, Ervin, 11, 28
Lee, Moira, 101, 102
Lehman, Rosemary, 73
Lerner, Richard, 61, 69
Leroy, H., 42
Letham, Robert, 150
Lightfoot, J. B., 152
Lincoln, Andrew T., 151, 219
Lock, Jennifer, 84, 102
Lohfink, Gerhard, 181, 182, 183
Louw, Johannes P., 13, 34, 43, 163, 180, 222
Lowe, Mary E., 101
Luo, Yiqi, 172
Lynch, James J., 133
Lytle, Julie, 70
Macaskill, Grant, 139, 150, 153
Maddix, Mark, 110
Magnusson, David, 61, 69
Malina, Bruce J., 208
Marchal, Joseph A., 151, 152
Margalef, R., 172
Marshall, Peter, 156
Mattis, Jacqueline S., 207
McGourty, G. T., 30
McGrath, Brendan, 138, 142, 143
McKenzie, Roderick D., 121, 126
Meadows, Philip, 104, 107, 108
Meyer, David, 106
Milgram, Stanley, 122
Milgrom, Jacob, 177
Milne, Esther, 73
Mitchell, Margaret M., 110, 111
Moll, Rob, 86
Moller, Leslie, 101
Montague, George T., 198, 201
Moo, Douglas J., 195
Moon, J. Robin, 133

Moore, Michael G., 169
Moreno, Jacob, 122
Morris, Norval, 155
Motyer, J. Alec, 31
Mowinckel, S., 209
Mrvar, Andrej, 125
Murabito, Joanne, 128
Needham, Phil, 199
Nesse, Randolph, 133
Neumayer, Christina, 118
Nida, Eugene A., 13, 34, 43, 163, 180, 222
Oberlinner, L., 214
O'Brien, Brandon J., 208
O'Brien, Peter T., 154, 155, 157, 161, 208
Oden, Thomas C., 74
Ogereau, Julien M., 155, 157
Ohmart, J., 30
Oldenburg, Ray, 81
Page, Charles R., II, 192
Palka, John, 92, 93
Palloff, Rena, 83, 84, 85, 90
Pannenberg, Wolfhart, 21, 22
Patten, Bernard C., 105, 106, 172, 188
Patterson, Elizabeth, 85
Pedahzur, Ami, 123
Pedersen, Johs, 209
Perliger, Arie, 123
Perriman, Andrew, 209
Perrin, Andrew, 94
Pfitzner, Victor C., 158, 159, 164
Pratt, Keith, 83, 84, 85, 90
Prensky, Marc, 114
Putnam, Robert, 67, 68, 76, 81, 83
Raffl, Celina, 118
Raptis, Dimitrios, 105, 106
Raymond, Jonathan S., 200
Reid, Marty L., 184
Reimer, Kevin S.,182, 186
Reinhartz, Adele, 161
Rheingold, Howard, 75, 79
Rice, Jesse, 77, 78
Richards, E. Randolph, 208
Ridderbos, Herman, 52, 141
Robertson, A. T., 180
Robinson, H. Wheeler, 209
Robinson, John A. T., 52, 59, 141

Roebuck, Carl, 54, 55
Rogers, Everett M., 129, 130, 189, 220
Rogers, Paul, 45
Rogerson, J., 209
Rosenqust, J. Niels, 128
Rothman, David J., 155
Rowen, Samuel F., 17, 80
Ryan, Edward J., 14
Sampley, J. Paul, 151, 152
Samra, James G., 196, 201
Sanders, Boykin, 161
Santmire, H. Paul, 25, 26
Scarfe, Adam, 69
Schaeffer, Francis A., 22
Schneck, Daniel J., 53
Schnelle, Udo, 55, 141
Schubert, Phil, 113
Schwarz, Christian, 60
Scott, James W., 47
Shanks, Hershel, 190
Shults, F. LeRon, 51, 52
Simard, Suzanne W., 48, 137, 138
Sinclair, A. R. E., 172
Smith, Marc, 75, 84
Smith, Robert Leo, 14, 29, 46

Smuts, Jan Christian, 172, 217
Snyder, Howard A., 5, 22, 27, 39, 51, 56, 60, 207, 210, 211
Son, Sang-Won, 151
Song, Felicia Wu, 117, 118
Stadd, Allison, 114
Stark, Rodney, 7, 194
Stavredes, Tina, 100
Stettler, Hanna, 191
Stoddard, John L., 217
Tan, Kim-Huat, 47
Tannehill, Robert C., 141, 150
Tansley, Alfred George, 14
Taylor, Vincent, 146
Taylor, Walter P., 209
Thayer, Joseph Henry, 163, 180
Thielman, Frank, 157, 158
Thompson, Clive, 76, 129
Thompson, James W., 167, 168
Tillich, Paul, 21
Travers, Jeffrey, 122
Utz, Rebecca L., 133
van Aarde, Timothy A., 57
VanderKam, James, 189
Volf, Miroslav, 208
Von Rad, Gerhard, 20, 21
Vygotsky, Lev, 61

Walsh, Carey Ellen, 31
Waltke, Bruce K., 19, 34
Ward, Richard F., 111
Ward, Ted W., 17, 80
Wasserman, Stanley, 220
Wayne, David, 86
Weng, Ensheng, 172
White, Ken, 91
Whitehead, Alfred, 68, 69, 121
Whitney, Donald S., 131
Wiles, Gordon P., 158, 159, 223
Wilkinson, Jennifer, 192
Willis, Avery T., Jr., 208
Wimberley, Edward T., 12, 13, 211
Worster, Donald, 33, 56, 121, 213
Wortman, Camille B., 133
Wright, N. T., 52, 139, 145, 189, 209
Yamauchi, Edwin M., 27
Yang, Yuanhe, 172
Yorke, Gosnell L. O. R., 52
Zerbe, Gordon, 24, 25

SUBJECT INDEX

absence, absent, 1, 67, 74, 111, 112, 159, 219
addicted, 114
agōn, 51, 158, 159, 164
See also ecological: motif; motif
alienation, 21-23, 78, 139
allēlōn, 7, 16, 42, 162, 170, 179-87, 198, 203, 217, 218, 224
Asclepius, 54
attachment, 96, 194
automatē, 43, 44
auxanō, 41, 42
barriers, 1, 74, 83, 118, 190, 204, 221
bidirectional, 71, 85, 90, 92, 95, 121, 126
biofeedback, 37, 53
body, 1, 7, 8, 38, 50, 52-55, 58-61, 83, 99, 112, 115, 131, 139, 141, 142, 146, 149, 150, 152, 156, 168, 181, 191, 197-200, 203, 204, 212, 223
  of Christ, 5-7, 16, 49-56, 58-63, 73, 77, 110, 119, 131, 137, 138, 141, 148, 150, 151, 156, 159, 166, 170, 171, 176, 179, 180, 182, 183, 187, 189, 195, 197, 199, 200, 202, 203, 205, 209, 210, 212, 216-18, 227
  human, 5, 6, 50-53, 55, 56, 60, 62, 212, 217, 227
  parts, 54, 55, 59, 60
bonds, bonded, 6, 15-17, 20, 21, 59, 73, 77, 87, 133, 138, 141, 143, 146, 167, 174, 177, 179, 200
Christ, 1, 5-7, 16, 18, 22, 23, 52, 55, 57-62, 73-75, 112, 113, 119, 131, 132, 134, 135, 137-50, 152,

153-58, 160, 161, 164, 166-72, 175, 176, 179, 180, 182, 183-86, 189, 193, 195, 196, 198-201, 203-5, 207-13, 216, 217, 218, 219, 221-25
  body of, 5-7, 16, 49-63, 73, 77, 110, 119, 131, 137, 138, 141, 148, 150, 151, 156, 159, 166, 170, 171, 176, 179, 180, 182, 183, 187, 189, 195, 197, 199, 200, 202, 203, 205, 209, 210, 212, 216, 217, 218, 227
classroom, 1, 2, 4, 6, 8, 72, 82, 83, 85, 88, 90-92, 99, 100, 102, 103, 107, 113, 148, 186, 205, 206
collaborate, collaboration, 2, 58, 93-97, 99, 100-102, 156, 158, 169, 208
communal, 70, 131
communion, 73, 75, 83, 151, 191, 219
  of the saints, 73, 74
community, 1-4, 6, 8, 13, 14, 19, 24, 28-30, 47, 51, 54-57, 61, 62, 67-86, 91-93, 95, 97, 99, 101-3, 106, 108, 113, 118, 121, 123, 124, 126, 131, 135, 147, 152, 160, 161, 163, 164, 166, 168, 169, 172, 175, 176, 179, 181-83, 185-87, 189, 196, 198, 199, 201, 202, 206-9, 211, 213, 220, 221, 223, 224
  physical, 79, 82
  virtual, 75, 79, 82, 97, 99, 110
conformed, 164, 167, 196, 201
connect, connection(s), 2-7, 11-13, 15-17, 20-22, 28, 29, 32, 35, 37, 40, 42, 52, 56, 59-63, 67-69, 73, 75-81, 83, 85, 86,

89, 90, 94-97, 102, 103, 105, 106, 108, 109, 114, 117, 119-35, 137-42, 144-50, 152-61, 163, 165-74, 179, 180, 182-84, 186-89, 193, 195, 197, 198, 200, 203-6, 210-14, 216-22, 225
  See also disconnection; ecological: connection; interconnection; social connection
contagion, 76, 128, 129, 131, 187-89, 193, 194, 198, 200, 205, 220
  gospel, 7, 194
  social, 15, 123, 125, 127, 128, 131, 188, 189, 194, 220, 222
  spiritual, 16, 188, 197, 200
contagious, 7, 15, 16, 125, 183, 188-93, 195-98, 202, 205, 222
  gospel, 193
  holiness, 189, 191, 193, 205
  sanctification, 195
contamination, 189-92, 202-4
context(s), contextualize, 26, 27, 33, 34, 36, 81, 84, 85, 90-93, 95, 101, 102, 106-10, 118, 122, 139, 146, 152, 155-57, 161-63, 173, 180, 195, 198, 206, 208, 212
cooperative, 157, 171, 176
corporate, corporately, 4, 56, 57, 60, 76, 77, 85, 106, 132, 151, 162, 168, 186, 196-202, 205, 209, 222-25
correlation, 30, 130
  See also interrelationships; relation(s)
course(s), 1, 2, 6, 8, 32, 40, 52, 69, 70, 77, 80, 81, 83-85, 88-91, 94, 97-102, 145, 168, 169, 186, 187, 195, 196, 206, 228

develop, development, developmental, 3, 7, 12, 16-18, 24, 26, 43, 61, 68, 69, 71, 79, 81, 82, 84-86, 89-91, 96, 98, 101, 102, 108-10, 112, 113, 115, 121-24, 126, 127, 130-33, 169, 173, 174, 177, 182, 185, 186, 199, 200, 207, 216, 221, 228

digital, digitalized, 3, 4, 8, 12, 67, 69, 70, 71, 76, 89, 91, 93, 95-97, 99, 104, 106-8, 113, 114, 116, 118, 124, 133, 148, 149, 206, 207, 221, 222

  ecology, 6, 35, 96, 103, 105-8, 110, 115, 118, 149

  hierarchies, 117

  immigrants, 98, 107, 114

  landscapes, 67, 105, 106, 113

  natives, 96, 104, 108, 114, 116, 117

  See also ecologies: digital

dimensions, 17, 18, 20, 75, 79, 107, 120, 132, 190

  six developmental, 18, 132, 185

disconnection, 5, 7, 21-23, 81, 139, 164, 167, 175

  See also connect

disembodied, 6, 79, 107, 113

distance(s), 1, 71, 72, 74, 75, 79, 83-85, 91, 92, 95, 101-3, 169

disturbances, 20, 45, 46, 126, 214

diversity, 19, 27, 28, 37, 46, 55, 56, 60, 97, 115, 125, 126, 181, 197

  biodiversity, 19, 46, 214

divine pedagogy, 82, 83, 112, 113

each other, 4, 6, 16, 17, 20-22, 40, 51, 53, 55, 60, 68, 69, 74, 75, 89, 101, 102, 105, 108, 114, 115, 124, 126, 127, 130, 133, 138, 148, 150, 159, 166, 167, 171, 172, 177-81, 185, 188, 197, 198, 204, 206, 212, 216-18, 220, 222, 225

ecological, 4-8, 12-14, 18, 20-22, 23-25, 27-37, 39, 40, 43-45, 47, 48, 49, 51, 53, 55, 56, 59, 60-63, 67, 68, 70, 71,

89, 104, 105, 108-10, 120, 121, 125, 133, 134, 137, 140, 146, 148, 160, 168, 169, 172-74, 180, 197, 200, 202, 203, 207, 209-14, 216, 221

  connection, 12, 13, 21, 42, 135, 166, 170, 203, 225

  interaction, 217

  motif, 4, 5, 25-27, 37, 227 (see also agōn)

  sanctification, 199, 200

  spiritual formation, 4, 137

ecologies, 4, 5, 7, 12, 13, 15-18, 20, 22, 27, 35, 62, 70, 86, 92, 95, 99, 105, 108, 110, 119, 126, 133, 168, 171, 174, 179, 183, 188, 189, 202, 204, 206, 211, 222, 227, 228

  digital, 6, 96, 97, 98, 103, 105, 106, 107, 108, 115, 118, 149

  learning, 6, 67, 86, 88, 91, 97, 98, 100, 103, 105, 149, 222

  See also ecological; ecology; ecosystem

ecology, 5, 7, 12-21, 24, 27-31, 33-34, 43-44, 46, 51, 56, 57, 63, 68, 89, 92, 105-7, 109, 114, 121, 132, 138, 145, 169, 172, 185, 188, 198, 204, 210, 216-17

  body of Christ, 6, 16, 60-62, 77, 131, 137-38, 148, 159, 170, 187, 189, 202, 212

  church, 6, 42, 51, 61, 137, 170-71, 195-96, 203, 204, 213

  cosmic, 12, 22, 30, 211

  digital, 105, 106, 110

  earth, 13

  environmental, 12

  faith, 70, 199, 225

  garden, 7, 12, 18, 19, 21

  human, 6, 16, 20, 50, 60, 62, 121, 126, 127, 227

  natural, 5, 20, 28, 33, 42, 56, 70, 88, 126, 171

  nested, 12, 13, 211

  oasis, 19

  online learning, 6, 87

  personal, 17, 132

salvation, 39

  of sin, 22, 27, 39, 51, 210

  social, 15, 20, 29, 61, 98, 121, 126, 173, 174, 200

soil, 44-48

spiritual, 6, 16, 17, 20, 42, 48, 60, 62, 77, 137, 138, 146-48, 157, 159, 169-72, 186, 195, 196, 202-4, 210, 212-14, 216, 217, 219, 221, 225

vineyard, 31, 33, 140

ecosystem, 4, 12-14, 17-20, 27, 28, 30-37, 45-48, 53, 56, 59, 60, 62, 68, 70, 71, 85, 88, 90, 92, 98, 103, 105, 106, 109, 110, 121, 125-27, 138, 160, 172-74, 179, 186, 188, 198, 202, 204, 207, 209, 211, 213, 214, 216, 217

  defined, 30, 37, 55, 56, 212, 213

  garden, 18, 20, 21

  human body, 17, 53, 55, 127, 173, 198

  See also ecological; ecologies; ecology; systems

edify, edification, 57, 61, 169, 170, 183, 184, 186, 198, 203, 225

empower, 61, 112, 113, 117, 119, 147, 148, 169, 181, 184, 186, 187, 193, 196, 202, 204, 205, 218, 220, 221

  See also power

environment(s), environmental, 6, 12-14, 18-21, 27, 29, 30, 36, 43-48, 61-63, 68, 69, 79, 81, 83, 84, 87, 92, 93, 97, 101, 106, 108, 109, 114, 118, 125, 126, 132, 133, 147, 148, 172, 174, 186-88, 192, 196, 200, 211, 214, 221

  See also habitat

exchange, 1, 3, 7, 28, 32, 35, 37, 42, 53, 59, 60, 61, 62, 69, 73, 75, 76, 91, 95, 102, 108, 109, 111, 119, 131, 134, 138, 156, 158, 172-74, 186, 205, 213, 214, 217-20

feedback, 2, 16, 73, 85, 172, 173

  See also biofeedback

fellowship, 7, 20, 73-75, 78,
104, 107, 108, 131, 139, 142,
148, 160, 176, 177, 181, 182,
186, 200, 201, 213, 218-22
table, 74, 110, 190, 191
field(s), 4, 5, 7, 11, 13-15, 26,
29, 30, 33, 34, 43, 44, 46-48,
56, 57, 67, 82, 122, 138, 148,
178, 200, 204, 215, 216, 227
flowers, 5, 33, 34, 37, 39, 42,
45, 52, 214-16, 227
formation, formative, 2, 4-8,
12, 18, 51, 65, 67, 70-74, 76,
77, 79, 84-86, 92-94, 97, 101,
102, 105, 109, 110, 112, 113, 115,
118-20, 123, 127, 130, 132, 133,
137, 159, 164, 166-68, 171,
181-85, 196, 199, 207, 210,
218, 220, 221
See also spiritual:
formation;
transformation
fragmented, fragmentation,
21, 78, 121, 139, 167, 175
garden, 7, 12, 20, 21
of Eden, 5, 13, 18, 19, 21,
22, 139
See also ecology: garden
Gentiles, 57, 58, 59, 61, 143,
151, 183, 189, 192, 195, 196,
218
grow, growth, 2-8, 12, 13,
15-19, 21, 27-52, 55-63, 68-71,
73, 74, 76, 77, 79, 81-86, 90,
92, 93, 95, 96, 102, 108-10,
115, 116, 119, 121, 122, 124,
126, 127, 130-32, 134, 137,
138, 140, 146, 148, 149, 154,
157, 159, 168-73, 175, 176, 179,
183-87, 194, 196-99, 201, 202,
204, 207, 209, 210, 212,
214-17, 220, 221, 223, 225,
227, 228
parables of, 40, 41, 48, 52,
214
habitats, habitation, 4, 13, 14,
28, 30, 47, 108, 109, 145, 212
See also environment;
inhabit
hesed, 37, 177, 178, 179
holistic, 2, 8, 24, 95, 132
holy, 57, 58, 139, 144, 166, 181,

184, 185, 189, 190, 193, 195,
196, 199, 200, 201, 203, 205,
224, 225
Holy Spirit, 32, 61, 73, 74, 75,
112, 113, 119, 147, 148, 152,
176, 181, 186, 187, 193, 197,
199, 200-202, 204, 205, 219,
221
homophily, 125
horticulture, 210, 225
hub, 124, 125, 129
imitate, imitators, 130, 152,
161, 162, 220
incarnate, incarnation,
incarnational, 75, 104, 107,
112, 113, 167
incorporate, 90
See also corporate
individuals, individualism,
individualized, 4, 5, 7, 16, 17,
20, 22, 46, 51, 53-57, 59-63,
67-69, 76, 77-80, 84, 85,
88-91, 94-96, 99, 100, 105,
107, 109, 112, 121, 122, 124-27,
129, 131, 141, 160, 168, 169,
172, 173, 181, 182, 186, 195-97,
199, 200-205, 208-10, 212,
214, 216, 217, 223-25, 227
influence(s), 6, 15, 16, 22, 54,
70-72, 75, 76, 81, 84, 86, 90,
92, 93, 102, 106, 109, 112, 119,
121, 123-29, 131, 133, 134, 161,
170, 172-74, 176, 183, 184, 188,
189, 194, 195, 197, 203, 207-9,
212, 217, 220, 221
inhabit, 14, 15, 33, 39, 57, 67,
71, 87, 115, 126, 133, 140, 196,
214, 222
See also habitat
intentional, intentionality, 47,
62, 85, 86, 90, 91, 94, 95, 97,
102, 103, 111, 178, 186
interaction(s), 1, 3, 4, 6, 7, 12,
13, 15-17, 19-21, 31, 33, 42, 53,
60-63, 68-71, 76, 77, 79, 80,
85, 86, 89-93, 95, 98, 101,
103, 106, 108, 114, 116, 122-24,
126, 127, 131, 133, 134, 137,
138, 140, 148, 168-77, 179,
181, 183-87, 195, 197, 198,
200, 201, 203-6, 212, 214, 217,
218, 220, 221, 225

intercession, 157, 158, 159, 166,
205, 223
interconnection,
interconnectedness, 3, 4, 6, 8,
11-13, 15-17, 20, 22, 27, 28,
30-33, 37, 48, 52, 55, 56,
58-60, 62, 63, 69, 70, 78, 88,
89, 96, 105, 115, 126, 127, 131,
133, 134, 138, 158, 160, 168, 169,
172-74, 180, 188, 198-200, 211,
212, 216, 217, 227
See also connect;
disconnection;
ecological: connection;
reconnect; social
connections
interdependence, 13, 14, 17, 28,
55, 56, 60, 69, 71, 121, 126,
212, 213
internet, 3, 70, 71, 79, 81, 83,
84, 86, 94, 96, 99, 104, 106,
114, 116, 117, 118, 120, 123, 221
interpersonal, 20, 21, 70, 79,
90, 100, 101, 109, 123, 133,
150, 167, 194, 213, 214, 224
See also ecology: personal
interrelationship, 27, 28, 37,
122, 180
See also correlation;
relation(s)
Jews, 58, 59, 61, 183, 189, 191,
195, 196
koinōnia, 74, 75, 152, 154-57,
160, 200, 219
learning connections, 169
learning interactions, 6, 91, 169
See also interaction(s)
letters, 6, 67, 73, 110-13, 137,
144, 147, 151-54, 156-58, 161,
162, 165, 167, 181, 184, 194,
218, 219, 224
maturity, maturation, 18, 33,
36, 37, 43, 47, 62, 63, 77, 92,
110, 146, 147, 159, 183, 184,
186, 197, 201, 204, 205,
207-9, 216-19
millennials, 91, 113-16
mission, 57, 104, 107, 108, 120,
121, 159, 164-66, 168, 169, 182,
183, 192, 202, 205, 208, 228
of the church, 57, 205
of God, 58, 182, 227

missional, 57, 58, 126, 194, 204, 208, 210
motif, 4, 5, 25, 26, 27, 37, 141-43, 152, 154, 156-59, 164, 227
  literary, 26
  See also *agōn*
mutual, mutuality, 4, 6-13, 16, 17, 22, 30-33, 35, 37, 42, 47, 53, 55, 56, 60, 61, 68, 69, 82, 88, 97, 98, 129, 131-34, 138, 155, 156, 158, 161, 166-68, 171-80, 183, 184, 186, 197, 198, 212, 213, 216, 218-20, 222
networks, 2, 3, 4, 6, 7, 13, 15, 53, 60, 67, 68, 70, 71, 76, 80-83, 85, 90, 93-96, 98, 99, 102, 103, 105, 106, 114, 119, 122-33, 137, 138, 158, 168, 169, 172, 188, 189, 194, 195, 207, 210, 220, 221
  See also social: networks
nourishment, 20, 29, 32, 35, 60, 62, 197, 204
nurture, 17, 83, 91, 107, 109, 110, 174, 200, 213
nutrients, 20, 29, 33, 35-37, 42, 43, 45, 47, 48, 53, 60, 62, 77, 108, 138, 140, 148, 154, 159, 171, 172, 186, 198, 203-5, 213, 214, 217, 218
offline, 75, 80, 115, 118, 124, 221
oikodomeō, 13, 57, 198
oikos/oikia, 12, 13, 14, 57, 58, 198
one another, 7, 15, 16, 18, 20, 27, 37, 42, 52, 55, 56, 61, 69, 70, 72, 74, 78-81, 89, 91, 97, 98, 100-103, 106-8, 110, 114, 119, 123, 127, 129-31, 134, 138, 148, 149, 152-54, 156-60, 162, 165-70, 176, 179-89, 198-200, 203-6, 212, 213, 218, 219, 221, 222, 224
organic, 30-33, 40, 42, 52, 56, 60, 62, 101, 138, 140, 146, 169, 171, 183, 186, 195, 210, 212, 214
organisms, 14, 33, 34, 37, 46-48, 51, 55, 56, 59, 62, 68, 88, 122, 137, 138, 146, 171, 172, 188, 198, 204, 209, 212, 216, 217
outcome, 106, 126, 134, 186
  targeted, 184, 185

parables, 4, 5, 35, 38, 39, 40-49, 53, 56, 62, 112, 131, 186, 187, 210, 214-16, 227
participate, participation, 1, 2, 48, 76, 83, 88, 96, 101, 102, 115, 116, 118, 119, 123, 133, 139, 143, 154-57, 164, 166, 167, 190, 223, 228
partner(s), partnership, 76, 152, 154-57, 159, 161, 163, 165, 169, 170, 177, 179
perfect, perfection, 5, 18-21, 33, 39, 46, 50, 58, 62, 68, 132, 139, 145, 171, 196, 198, 222-24
plural, plurality, 78, 159, 196, 201, 222-24
power, powerful, 5, 6, 11, 15, 17, 30, 46, 51, 58, 59, 73, 74, 76, 110, 119, 122, 127, 128, 133, 143, 144, 148, 152, 168, 170, 174, 178, 188-91, 193, 203-6, 216, 220, 222, 223
  See also empower
prayer, 7, 77, 88, 90, 98, 115, 117, 147, 148, 157-59, 166, 181, 199, 205, 221, 223, 224, 227, 228
prepositions, prepositional, 29, 30, 34, 44, 45, 59, 138, 140-42, 144, 148, 152, 153, 157, 188, 216
presence, 6, 36, 39, 45, 70, 73-75, 77, 85, 100, 110-13, 115, 173, 181, 192, 197, 204, 224
  mediated, 72, 110, 111, 113
  virtual, 110
priests, priesthood, 166, 176, 193
process, 2, 5, 17, 29, 34, 38-41, 43, 44, 50, 52, 56, 62, 68, 70, 83-86, 88-91, 93, 97, 98, 100-102, 105, 109, 112, 130, 159, 172, 173, 184-86, 192, 195, 201, 203, 205, 207, 208, 215, 216, 218, 222, 225
reciprocal, 7, 16-18, 32, 33, 37, 42, 60-62, 67, 69, 81, 91, 93, 95, 98, 101, 102, 110, 119, 122, 126, 131, 133, 140, 148, 158, 159, 170, 172-81, 183, 184-87, 198, 204, 217-19
  community, 181-83

reciprocity, 16, 20, 68, 69, 75, 76, 80, 84, 97, 101-3, 109, 118, 123, 126, 127, 130, 132, 165, 173, 177, 182, 219, 220
reconcile, reconciliation, 22, 52, 62, 111, 139, 146, 163, 164, 208, 211
reconnect, 78, 139, 140, 204
  See also connect
relation(s), relationships, relational, 4, 7, 14-16, 20, 21, 27, 30, 32, 33, 35, 37, 40, 41, 44, 45, 52, 55, 56, 58, 61, 62, 67-70, 74, 75, 78, 79, 80, 82-85, 90, 91, 93, 95, 96, 98, 100-103, 105, 107-11, 114, 121, 123, 124, 127, 129-33, 139-41, 143, 146, 147, 149-51, 154, 156-58, 162, 163, 165, 167-72, 175-84, 186, 188, 194-96, 199-201, 203, 204, 209, 212-14, 216-19, 224
resilience, 46
restore, restoration, 23, 33, 36, 62, 164, 203
salvation, 22, 23, 27, 39, 40, 48, 51, 58, 140, 142, 143, 147, 150, 169, 182, 192, 193, 199, 200, 207, 208, 210
sanctification, 7, 58, 131, 147, 186, 191, 195, 196, 199-202, 223, 224
Scripture, 5, 26, 27, 29, 36, 42, 45, 60, 115, 126, 150, 175, 185, 193, 208-11, 220, 225
Second Life, 115
seeds, 5, 34-40, 42-48, 52, 214-16, 227
  mustard, 40, 42, 47, 48
separation, 15, 16, 21, 32, 128, 129, 193
  six degrees of, 15, 16, 122
shalom, 21
social, 15
  connections, 6, 16, 17, 20, 22, 81, 90, 96, 103, 123-26, 128, 130, 131, 133, 186, 188, 221, 222
  networks, 2, 6, 7, 15, 53, 67, 68, 71, 76, 81, 82, 93-95, 103, 122-33, 188, 189, 194, 195, 220

See also under contagion; ecology

soil, 29, 34, 35, 42-48, 108, 126, 214, 216

sōma, 52, 54, 141

spiritual, 2-8, 12, 15-18, 20, 21, 22, 25, 26, 28-33, 35-37, 39-42, 44, 48, 51-53, 57-63, 65, 71, 73-75, 77, 79, 82, 85, 86, 88, 91, 101, 107, 108, 110, 112, 115-17, 126, 130-34, 137-40, 143, 146-50, 154-60, 166, 167, 169-73, 182-90, 192, 195-205, 207-10, 212-22, 225, 227, 228

   formation, 4, 5-8, 18, 65, 67, 85, 86, 92-94, 101, 110, 112, 113, 118, 120, 127, 130, 133, 137, 159, 164, 166, 168, 171, 184, 185, 196, 207, 210, 218, 220, 221

   See also under contagion; ecological; ecology

strategies, 89, 91-93, 95, 97, 98, 100, 204, 228

syn, 58, 59, 60, 138, 140-42, 144, 145, 148, 151-54, 156, 158, 160, 163-65, 169, 216

   See also *syn*-compounds

syn Christō, 52, 141, 142, 146, 148, 153

syn-compounds, 7, 16, 58, 139, 141-43, 145, 146, 148, 150-65, 183, 184, 197, 216-18

   horizontal, 7, 149-51, 154, 157, 159, 160, 162, 163, 165, 170, 171

   vertical, 6, 142, 143, 146, 150, 160

systems, 3, 13, 14, 20, 50, 53, 68, 70, 89-92, 102, 103, 105, 106, 108-10, 120, 121, 126, 169, 172, 203, 210, 213

   See also ecosystem

third place, 81, 82

together, togetherness, 13, 16-18, 20, 52, 57-60, 62,

67-71, 73, 75-77, 79, 81-86, 89, 91, 93, 95, 96, 98, 100, 102, 104, 108, 110, 115, 117, 125, 130, 132, 137, 138, 140, 144-46, 148, 151, 154, 156, 158-67, 171, 175, 178, 179, 181, 182, 184, 185, 196, 198, 204, 205, 209, 216, 218, 219, 223

transformation, transformative, 82, 84, 85, 107, 108, 113, 119, 126, 131, 132, 164, 167, 168, 184, 186, 219, 220

   See also ecological: spiritual formation; formation; spiritual: formation

tree, 29, 39, 48, 68, 138, 195

vine, vineyards, viticulture, 4, 5, 30-33, 35-37, 39, 140, 146, 210, 216, 225, 227

whole person, 18, 20, 131, 132, 160, 184, 186, 219, 228

SCRIPTURE INDEX

OLD TESTAMENT

Genesis
1, *139, 145*
1:26-28, *78*
2, *19*
2:5, *19*
2:6, *19*
2:7, *20*
2:8, *19*
2:9, *19*
2:10-14, *19*
2:15, *19, 20*
2:18, *15*
2:18-25, *20*
2:19-20, *20*
2:24, *20*
3, *21*
3:8, *20*
3:12, *22*
3:17, *22*
5:1-2, *78*
12, *145*
12:3, *126*
13, *19*
13:10, *19*
15, *223*
17, *223*

Numbers
15:20-21, *195*
22, *39*
22:2, *39*
22:5, *39*
22:31, *39*
23:9, *39*
24:3-4, *39*

Deuteronomy
18:4, *195*
26:17-18, *177*

Joshua
7, *202*

Ruth
1:16, *178*
4, *178*
4:13, *178*
4:17, *178*

1 Kings
21, *31*

Job
38, *28*

Psalms
1:1, *29*
1:1-3, *29*
1:3, *29*
13, *36*
22, *36*
30, *36*
40, *36*
42, *32*
65:9-13, *36*
80, *32*
80:8, *32*
80:8-11, *31*
80:8-16, *30, 31*
80:9-11, *32*
86, *36*
90, *28*
103:15, *33*
103:15-16, *33*
104, *28*
104:14-15, *28*
104:16-17, *28*
104:16-23, *28*

Proverbs
27:17, *61, 183*

Song of Solomon
2:2, *33*

Isaiah
5, *32*

5:1, *31*
5:1-7, *30, 31*
5:2, *31*
5:5, *30, 31*
5:6-7, *32*
5:7, *30, 31*
30:23, *34, 35*
35:1, *33*
47:15, *175*
53:6, *175*
56:11, *175*

Jeremiah
2:21, *30, 31*
6:9, *31*
17:7-8, *29*
31, *179*
31:5, *31*

Ezekiel
17, *35*
17:5-6, *34, 35*
17:7, *36*
17:7-10, *30*
17:9-10, *36*
17:20-21, *36*
31:4-5, *29*

Hosea
14:5, *33*
14:5-7, *33*

Joel
2:21-22, *36*

Amos
9:13-15, *30, 36*
9:14, *31*

Zephaniah
1:13, *31*

NEW TESTAMENT

Matthew
6:28, *33, 34, 215*

8:14, *57*
13, *35*
13:24, *44*
13:31, *47*
13:31-32, *34, 41*
13:32, *47*
20:1-11, *31*
21:33-43, *31*
27:44, *143*
28:16-20, *182*
28:19, *205*

Mark
1:41, *191*
4, *35, 192*
4:1-12, *42*
4:8, *34, 41*
4:26-27, *43*
4:26-29, *34, 42*
4:27, *43*
4:28, *43*
4:30-34, *42*
4:35, *192*
4:41, *47*
5, *191, 205*
5:1, *191, 192*
5:11, *192*
5:20, *192*
5:21, *192*
5:25-34, *191*
5:41, *191*
9:2, *167*
11:13, *39*
14:48, *163*
15:32, *143*

Luke
1:80, *41*
2:32, *192*
2:40, *41*
2:52, *132*
5, *190, 191*
5:7, *163*
5:29-30, *190*

5:30, *190*
5:31, *191*
7:14, *191*
8:5, *46*
8:5-8, *44*
8:6, *45, 46*
8:7, *46*
8:8, *45, 46*
12:32, *182*
12:54, *39*
13:19, *41, 47*
22:20, *179*
23:12, *179*
24:14, *179*
24:17, *179*
24:32, *179*
24:51, *145*

John
3:16, *139*
4:35, *39*
13:14, *183*
13:34, *180, 183*
14, *48*
14:4, *32*
14:12, *32*
14:16, *32*
14:18, *32*
14:25-26, *32*
15, *31, 140*
15:1-11, *30, 32*
15:4-7, *33*
16:13, *201*
19:32, *143*

Acts
1:11, *145*
2, *208*
2:1-4, *202*
2:6, *208*
2:8, *208*
4:8, *202*
4:31, *202*
6:7, *193*
9:1, *141*
9:4-5, *141*
10:2, *57*
10:22, *57*
10:30, *57*
10:44, *202*
11:15, *202*
12:3, *163*
12:10, *43*

12:24, *193*
13:49, *193*
13:52, *202*
15:39, *179*
19:6, *202*
19:20, *193*
20:28, *131*

Romans
1:1, *184*
1:2, *193*
1:5, *147*
1:8-15, *184*
1:11-12, *184*
1:12, *184, 218*
6, *143*
6:3-5, *150*
6:4, *140, 144, 147*
6:4-8, *140*
6:5, *140, 146, 148*
6:6, *140, 143, 146, 148*
6:8, *140, 141, 148*
6:11, *146*
7:4, *55, 148*
8:17, *144, 145, 148*
8:18-21, *145*
8:23, *195*
8:28, *18*
8:32, *140*
8:37-39, *140*
8:39, *148*
9, *195*
9:14-18, *196*
11:16, *195*
11:16-24, *195*
12, *55, 184*
12:1, *196*
12:2, *167*
12:4-5, *55*
12:5, *61*
12:6, *61*
12:10, *184*
12:16, *218*
13:8, *184*
14:19, *57, 184*
15:2, *57*
15:14, *184*
15:30, *158*
16, *165*
16:5, *57*
16:16, *181, 184*
16:26, *147*

1 Corinthians
1:9, *74, 200*
1:10, *203*
1:10-11, *167*
1:10-13, *152*
3:1, *203*
3:1-4, *62, 77, 152*
3:3, *203*
3:6, *28, 29, 43*
3:6-9, *204*
3:7, *215*
3:9, *44, 57*
4:1-2, *57*
4:16, *161*
5, *202*
5:1, *202*
5:2, *202*
5:6, *202*
7:5, *179*
8:13, *203*
9:23, *154, 155*
9:24-27, *164*
10:16, *75*
10:16-17, *55*
10:24, *203*
10:32-33, *203*
11:1, *161*
11:17-19, *152*
11:30, *203*
11:33, *203*
12, *51, 55, 131, 203, 217*
12:7, *61*
12:11, *61, 208*
12:12, *131*
12:15-21, *55*
12:20, *55*
12:24-26, *218*
12:25, *61, 203*
12:27, *52*
14:1, *203*
14:3, *57*
14:5, *57*
14:12, *57, 61, 203*
14:26, *57*
15:20, *195*
15:23, *195*
15:26, *144*
15:55-56, *144*
15:57, *144*
16:19, *57*
16:20, *203*

2 Corinthians
1:8, *158*
1:8-11, *158*
1:11, *158*
2:1-4, *111*
3:1-11, *222*
3:18, *167, 196*
4:14, *141*
5:17, *52, 205*
5:18-20, *22*
5:18-21, *163*
5:21, *201*
7:1, *196, 222, 223*
8:14, *219*
8:16-17, *111*
12:20-21, *111*
13:4, *140*
13:9, *158*
13:12, *180*
13:14, *75, 201, 219*

Galatians
2:20, *143*
3:6-14, *223*
3:18, *145*
3:29, *145*
4:1, *145*
4:2, *57*
4:7, *145*
4:19, *167*
4:30, *145*
5:13, *180*
5:21, *145*
5:22, *218*
6:6, *219*
6:6-10, *146*
6:10, *57*

Ephesians
1, *147*
1:4, *196*
1:10, *57, 58, 211*
1:11, *145*
1:14, *145*
1:16, *158*
1:18, *145*
1:22-23, *55*
2, *57*
2:4, *145*
2:5-6, *144*
2:6, *140, 145, 147*
2:15, *58*

2:18, *58*
2:19, *57, 131*
2:19-21, *146*
2:19-22, *57, 58*
2:20, *58*
2:21, *41, 57, 131, 196, 198*
2:22, *58*
3:2, *57, 58*
3:4, *58*
3:6, *58*
3:14-21, *147*
3:16, *148*
3:20, *148*
4, *7, 55, 131, 147, 148, 217*
4:1-16, *51, 197*
4:2, *183*
4:2-3, *218*
4:3, *77*
4:3-4, *197*
4:4, *197*
4:11-16, *48*
4:12, *55, 57*
4:13, *18, 132, 185*
4:15, *18, 185*
4:15-16, *197*
4:16, *52, 55, 57, 59, 61, 154, 198*
4:22, *200*
4:24, *200*
4:25, *183*
4:29, *57*
4:32, *180, 183*
5:1, *161*
5:21, *183*
5:23, *55*
5:26-27, *225*

Philippians
1:1, *152*
1:3-7, *166*
1:4, *152, 158*
1:5, *152, 154, 155, 159*
1:6, *222*
1:7, *152, 154, 155, 156, 157, 159, 164*
1:8, *152*
1:9-11, *157*
1:12, *154*
1:12-20, *166*
1:18-29, *164*
1:19, *157*
1:22, *152*

1:23, *141, 142, 152*
1:25, *152*
1:27, *152, 159, 160, 164*
1:28-30, *166*
1:30, *159*
2, *159*
2:1, *142, 152, 200, 219*
2:2, *152, 160, 164*
2:2-3, *218*
2:2-4, *162*
2:3, *152*
2:5, *164*
2:5-11, *166*
2:6, *167*
2:6-7, *167*
2:7, *167*
2:9-11, *145*
2:12, *147, 166*
2:17, *152, 166*
2:17-18, *152, 218*
2:18, *152*
2:19, *160*
2:19-30, *166*
2:20, *160*
2:25, *152*
2:30, *160*
3:5-16, *165*
3:7-16, *164*
3:8, *161, 164*
3:10, *152, 164, 167*
3:13, *164*
3:14, *161, 164*
3:15, *164*
3:15-16, *161*
3:17, *152, 160, 162*
3:19, *164*
3:21, *152, 167*
4, *155*
4:1, *162*
4:2, *152, 162, 164, 213*
4:3, *152, 159, 162, 163, 164*
4:10, *164*
4:14, *152, 154, 155, 156, 165*
4:15, *152, 155, 156, 165, 219*
4:15-16, *152*
4:15-18, *166*
4:17, *156*
4:17-18, *165, 166*
4:18-19, *167*

Colossians
1:3, *158, 218*
1:9, *158*
1:18, *55*
1:21-22, *139*
1:24, *55*
1:28, *201, 219, 225*
2:5, *1*
2:10, *225*
2:12, *140*
2:13, *140*
2:14, *143*
2:19, *42, 55, 60, 61, 62, 176*
2:20, *141*
3:1, *146*
3:1-2, *147*
3:3, *141*
3:4, *140*
3:13-14, *218*
3:16, *180*
4:3, *218*
4:15, *57*

1 Thessalonians
1:1-2, *224*
1:2, *158*
1:6, *161*
2:8, *224*
2:14, *161*
2:19-20, *224*
3:9-10, *224*
3:11-13, *223, 224*
3:12, *224*
4:1, *224*
4:3, *224*
4:3-4, *224*
4:6, *224*
4:7, *196*
4:9, *224*
4:9-12, *224*
4:11, *224*
4:12, *224*
4:13, *144*
4:14, *140*
4:17, *141*
4:18, *180, 224*
5:6, *224*
5:10, *140*
5:11, *57, 180, 181, 224*
5:15, *224*
5:22-23, *224*
5:23, *224, 225*
5:23-24, *223, 224*
5:25, *158*

2 Thessalonians
1:3, *184*
3:1, *193*
3:1-2, *158*
3:7, *161*
3:9, *161*

2 Timothy
2:11, *140*
2:12, *140*
3:5, *167*

Hebrews
6:20, *166*
10:1-25, *166*
10:11-18, *179*
10:12, *145*
10:19, *179*
10:24, *179, 180, 184*
13:15-16, *166*

James
1:4, *225*
1:18, *195*
5:16, *159, 180*

1 Peter
2:1, *214*
2:1-2, *213*
2:2, *42, 214*
2:4-10, *166*
4:9, *181*
4:10, *57*

2 Peter
3:18, *42, 140*

1 John
1:3, *75*
1:3-7, *200*
1:6, *75*
1:9, *141*
4:4, *204*
5:18, *204*
5:19, *204*

2 John
5, *213*

Revelation
7:9, *126, 208*
14:4, *195*

Finding the Textbook You Need

The IVP Academic Textbook Selector
is an online tool for instantly finding the IVP books
suitable for over 250 courses across 24 disciplines.

ivpacademic.com